KU-161-457

UNDERSTANDING URBAN UNREST

From Reverend King to Rodney King

DENNIS E. GALE

SAGE Publications
International Educational and Professional Publisher
Thousand Oaks London New Delhi

Copyright © 1996 by Sage Publications, Inc.

All rights reserved. No part of this book may be reproduced or utilized in any form or by any means, electronic or mechanical, including photocopying, recording, or by any information storage and retrieval system, without permission in writing from the publisher.

For information address:

SAGE Publications, Inc.
2455 Teller Road
Thousand Oaks, California 91320
E-mail: order@sagepub.com

SAGE Publications Ltd.
6 Bonhill Street
London EC2A 4PU
United Kingdom

SAGE Publications India Pvt. Ltd.
M-32 Market
Greater Kailash I
New Delhi 110 048 India

Printed in the United States of America

Library of Congress Cataloging-in-Publication Data

Gale, Dennis E.
    Understanding urban unrest: From Reverend King to Rodney King/
author Dennis E. Gale.
      p. cm.
    Includes bibliographical references and index.
    ISBN 0-7619-0094-2 (acid-free paper).—ISBN 0-7619-0095-0 (pbk.:
acid-free paper)
    1. Violence—Government policy—United States. 2. Urban poor—
Government policy—United States. 3. Riots—United States.
4. Urban policy—United States. 5. Community development, Urban—
United States 6. United States—Race relations. 7. United
States—Social conditions. I. Title.
HN90.V5G35 1996
303.6'2'091732—dc20   95-50225

This book is printed on acid-free paper.

96 97 98 99 00 10 9 8 7 6 5 4 3 2 1
Sage Cover Designer: Candice Harman
Sage Production Editor: Michéle Lingre

Contents

Preface

"Soul Brother," the sign read. It was nothing more than a piece of plywood hastily tacked up on a storefront with lettering in bright green paint. Only the day before, arson, looting, and vandalism had flared along Blue Hill Avenue in Boston's Roxbury section. White delivery people were chased, a few whites were pummeled by young blacks, and police officers ducked bottles and rocks from the angry crowd. Broken glass was everywhere, the smell of charred wood pervaded the area, and sodden debris from looted shops lay in the gutters. As I drove through the city's largest black neighborhood that day in 1967, my heart pounding, I was sure that mine was the only white face anywhere in sight. I averted my eyes from the gangs of young black men on the streets, but I noticed that the Soul Brother sign must have worked. It was attached to the only store on the block with no visible damage.

Just then I heard a warning: "Go home, Honkie. You don't b'long here." My foot pressed down on the accelerator.

A year later, I was back in Roxbury, this time as a summer intern at the Boston Redevelopment Authority, helping to write reports on various Urban Renewal projects. Our project site office was in an abandoned synagogue, a relic left behind by departing middle-class Jewish families, most of whom had moved to the suburbs. I became friends with a young Presbyterian minister, Don Campbell, who was on the model neighborhood board in Roxbury. Soon I was one of his volunteers. He introduced me to Bernie Frieden and Lang Keyes, both professors at Massachusetts Institute of Technology (MIT) who were affiliated with Boston's Model Cities program. I had plans to study urban policy at Harvard's graduate school in the fall and I talked to them about the city planning profession. They described MIT's curriculum, compared various university programs, and spun stories about their experiences working on Boston's inner-city problems. Before long, I was hooked.

Although he advised me to forget city planning and study sociology instead, Harvard Professor Nathan Glazer was an inspiration. At his request, I gave a short presentation in his graduate urban social policy class on Martin Anderson's (1964) *The Federal Bulldozer* and Boston's Urban Renewal program. I didn't realize it at the time, but I had just discovered my second love—teaching.

A fledgling scholar at Harvard, Doris Kearns (now Doris Kearns Goodwin) piqued my interest in the Johnson administration's War on Poverty. Then coteaching a graduate seminar, she shuttled each week to the White House while writing her seminal biography of Lyndon Johnson. Much to my regret, I missed Daniel Patrick Moynihan's graduate class in urban policy due to his departure from the university to become President Nixon's Urban Affairs Advisor. Alas, Edward Banfield's popular seminar in urban politics, cotaught with James Q. Wilson, also eluded me due to an enrollment cap. Nonetheless, Moynihan's (1969) *Maximum Feasible Misunderstanding* and Banfield's (1974) *The Unheavenly City Revisited* contributed to the development of this book.

During the summer of 1969, I served as an urban intern in the Model Cities Administration at the U.S. Department of Housing and Urban Development (HUD). The year before, Washington had been torn by its worst civil disturbance in half a century. That memory was still fresh in the minds of my HUD colleagues, who were only too aware that Model Cities was perceived as a riot antidote by many mayors and members of Congress.

A young consultant to HUD, Marshall Kaplan, would later coauthor an authoritative book with Bernie Frieden about the Model Cities demonstration. He and Assistant Secretary Floyd Hyde, a Nixon appointee and former mayor of Fresno, were two of the most knowledgeable Model Cities insiders at the time. I listened to everything they had to say.

By the time my internship ended, I had seen enough to know that this most unusual government program literally groaned under the weight of expectations it bore. Not yet a cynic about the program or about the future of American cities, I decided that my education had only just begun. I went on to study city planning at the University of Pennsylvania, where William Rafsky and Mort Schussheim, among others, prodded and tweaked my eager, although uncertain, intellect. Rafsky had served on the commission that first recommended the Model Cities demonstration to President Johnson. Schussheim (1974) would author *A Modest Commitment to Cities,* sharing his wisdom as a longtime analyst of urban policy in Washington.

Returning to Washington in 1971, I practiced city planning, carried out research at the Urban Institute, and commenced my career as an urban planning professor at George Washington University. It would be many years, however, before the opportunity would arise to reflect on my Model Cities experience. One important inspiration was my introduction to Robert C. Wood in 1991. Then the Luce Professor at Wesleyan College, he shared his experiences as undersecretary at HUD in the Johnson administration. Although it is only one of his impressive accomplishments, he can take a major share of the credit for midwiving the Model Cities program to life. He encouraged me to finish this study and offered several critical suggestions for revising the manuscript. However imperfectly I have followed through on his advice, I am indebted to him. Others who have read and helpfully commented on earlier drafts include Todd Swanstrom of the Rockefeller Institute at the University of Albany and three anonymous reviewers. Like Bob Wood, however, they bear no responsibility for anything I have written—or failed to write—in this book.

As the reader will see, I have borrowed shamelessly from the ideas and observations of these talented teachers and authors. That they touched my life at many points over the years is a measure of how blessed I feel. Likewise, I owe a debt of gratitude to librarians and collections at the Edmund S. Muskie Archives at Bates College, the Rotch and Dewey libraries at MIT, the Weidner, Houghton, and Lamont libraries at Harvard, the Gelman Library at George Washington University, the Library of Congress, the HUD library, and the Martin Luther King, Jr. Library of Washington, D.C. Finally, I thank Carrie Mullen, Renée Piernot, Harry Briggs, Jessica Crawford, Deidre McDonald Greene and their colleagues at Sage for their confidence in my manuscript and their suggestions for refining it. My colleagues and students at Florida Atlantic University have provided me with an engaging scholarly home in which to prune and refine the final drafts of this book. To them I am also grateful.

DENNIS E. GALE

1

Introduction

If the 1990s have brought us nothing else, the decade has been witness to the worst episode of urban rioting in modern American history. Born out of anger at the court's decision in the Rodney King beating incident, the violence erupted in Los Angeles on April 29, 1992, and quickly took on cataclysmic proportions. When the looting, burning, and beating were over, 50 people were dead, more than 2,000 were injured, and more than 1,000 properties were destroyed.

Perhaps the greatest tragedy associated with the Los Angeles riots, however, was their familiarity. American television viewers watched the evening news reports with uneasy recollection. We had seen it all before in the 1960s. But for many people, accompanying their dismay at the cruelty and destruction of the riots was a measure of sympathy for the anger and sense of betrayal displayed by those in the streets of South-Central Los Angeles. No matter how reprehensible the destruction and cruelty there, the jury's decision in the Rodney King trial begged reexamination.

This book is not about the Rodney King decision nor is it about the Los Angeles riots per se. It is inspired in part by those events and by other

1

similar tragedies preceding them. My purpose here is to examine how the
federal government responds to urban poverty and mob violence. How, in
other words, have Congress and the White House interpreted such events,
how have they misperceived them, and how has the character of their
response changed over time? If poverty is at the root of urban crowd
violence, how have policies to reduce the conditions that contribute to such
violence been formulated in Washington?

Many Americans are sadly aware that the violence in Los Angeles in
1992 was no isolated incident. It was only the most spectacular event in a
long line of urban riots that have erupted periodically during this century.
Most Americans are old enough to recall the worst succession of such
outbursts in the 1960s, which culminated in nationwide rioting after the
assassination of Martin Luther King, Jr., in 1968. Since then, scattered
incidents have flared in cities such as Miami and Washington. But even
before the wrenching 1960s, America had suffered major riots in East St.
Louis, Chicago, Detroit, Harlem, and several other cities. One is hard put
to name another country in which urban interracial mob violence has been
such a recurring theme. It is unlikely that this legacy will disappear from
the American landscape in the foreseeable future. Urban riots will occur
as long as there are large numbers of disaffected people living in tightly
concentrated enclaves of poverty and discrimination who witness ques-
tionable treatment of others like themselves. We can count on that.

There have been efforts to address the presumed underlying causes of
urban rioting in the past. The most remarkable involved a panoply of
federal programs cranked out of Congress in the mid- and late 1960s. One
in particular was designed to address a complicated labyrinth of urban
social problems. Called Model Cities, the legislation was enacted in the
closing days of the 89th Congress in 1966. Perhaps more so than any
federal initiative in history, Model Cities sought to grapple with the
problems feeding the unrest in America's poor urban neighborhoods. What
was learned from that noble experiment? To what extent, if any, did those
lessons influence federal policy development in the wake of the Los
Angeles riots of 1992? I examine these matters in this volume.

The central thesis of the book is the following: Over a span of 30 years,
the U.S. government has grappled with urban mob violence by funding
programs whose benefits are limited to poor neighborhoods. In doing so,
Congress has accepted the premise that alleviating the many dimensions

of poverty in these enclaves is the best preventative for urban unrest. Whatever the legitimacy of these views in the past, in the 1990s, and beyond, place-based policies bear very limited promise. The nation cannot alleviate either poverty or the penchant for mob violence through excessive reliance on programs such as Model Cities or Empowerment Zones-Enterprise Communities. The shifting problem context of social conditions and the vastly divergent political environment in which they are perceived today command new policy options. A profound transformation in the manner in which urban policy is framed has emerged in Washington. It represents perhaps the most critical turning point in the nation's uncomfortable relationship with its cities since the 1960s.

2

Los Angeles '92
Was Nothing New

Few events in modern times have so thoroughly galvanized Americans as the scenes conveyed by a shadowy home video recording of four Los Angeles police officers beating a black man senseless. An 81-second strip of raw brutality, the vignette has been viewed by more people perhaps than any visual recording since the film of the assassination of John F. Kennedy in 1963. The segment, shot early on the morning of March 3, 1991, shows Rodney King writhing on the roadside as white police officers appear to take turns beating him with batons. Their routine is broken up by occasional kicks delivered to King's body.

The videotaped scene was telecast locally the next day. Over the next 14 months, it would be replayed countless times in American homes during local and network news and documentary programs. Los Angeles police department recordings of radio messages from the scene of the incident reveal a callous attitude about the King beating from officers at the scene ("I've got a victim with . . . ha ha . . . head injuries."). Public outrage over the incident caused Los Angeles Mayor Tom Bradley to ask Police Chief

Daryl G. Gates to resign. Knowing that the mayor had no authority to fire him, Gates refused. By the time the trial of the four police officers began, almost 1 year later, the infamous video recording had become the equivalent of a nationwide visual incantation on the hidden realities of police victimization.

The Court Renders a Verdict

Many Americans took for granted the culpability of the four defendants. After all, who could question the act itself? It was recorded on video. We could see the event with our own eyes; seeing is believing, isn't it? At question in the minds of many observers was not whether a crime had been committed, but how severe the officers' sentences would be. But the California legal system insisted on a different standard of judgment. The verdict turned not on the fundamental notion that four armed people beating an unarmed and unresisting victim is a violation of the law. Instead, it was based on the conclusion that King had appeared to resist arrest and the officers were following prescribed police department procedures in subduing him. No excessive use of force had occurred, the jury ruled, and three of the four officers—Stacey Koon, Theodore Briseno, and Timothy Wind—were acquitted on Wednesday, April 29, 1992. A mistrial was declared in the case of Laurence Powell. All four officers thus escaped conviction. Adding to the public frustration over the decision was the fact that the trial was held in suburban Simi Valley, an overwhelmingly white middle-class community, home to many police officers and firefighters. Furthermore, the jury was made up of 6 men and 6 women, 11 of whom were whites and 1 of whom was Hispanic. Not a single black citizen participated in the jury's decision. Adding to public mistrust, the victim, Rodney King, was never allowed by his own attorney to testify during the trial.

Riots Erupt

As soon as word of the verdict got out, a tidal wave of anger welled up at the intersection of Florence and Normandie avenues in South-Central

Los Angeles. Soon, buildings were being torched, stores were being looted, and passing motorists were being pelted with missiles. Over in Lake View Terrace, an angry crowd rallied at the spot on the highway where King had been beaten a year earlier. They moved their demonstration to the police precinct station where the four acquitted officers were assigned (Serrano & Wilkinson, 1992).

By Thursday, April 30, National Guard troops had arrived and had begun to patrol key points in the riot sections of South-Central. The next day, Mayor Bradley declared a dusk-to-dawn curfew. Protesters called for a rally downtown. About 150 participated, protesting the police acquittal and denouncing Police Chief Gates. A few hundred people gathered in Koreatown, marching through parts of South-Central calling for peace. Around the riot areas, small bands of people wielded brooms and shovels, cleaning debris from streets and boarding up buildings ("Understanding the Riots," 1992).

On Friday evening, U.S. President George Bush ordered 4,500 Army and Marine troops to Los Angeles. By Sunday May 3, 3,500 National Guard troops were in the streets and another 3,200 were on reserve. Backing them up were 1,000 federal law enforcement officers, including border patrol guards and Federal Bureau of Investigation (FBI) agents. Sporadic acts of violence were occurring, but the numbers of fires, beatings, and lootings had abated considerably. Unremarkably, gun store owners in southern California reported a substantial increase in business in the days following the eruption of violence.

The full effect of the mayhem was not confined to Los Angeles. Up the Pacific coast in Seattle, the King verdict was followed by 2 nights of fires and 115 arrests. About 150 youths rampaged downtown, shattering store windows, assaulting bystanders, and starting several small fires. Authorities reported that most of those arrested were white and that minority neighborhoods were quiet. Other incidents of civil unrest occurred in San Francisco, Long Beach, and Madison, Wisconsin. In Atlanta, 370 arrests and 73 injuries, mostly not serious, were reported. Disorders associated with the King verdict broke out in Newark, Detroit, New York, Washington, Chicago, and Philadelphia—all scenes of rioting in the 1960s. Mercifully, these were relatively mild events and all dissipated quickly. Though nothing like the tumult in Los Angeles, these smaller riots spread fear that another 1960s-style succession of civil disturbances was in the offing. Meanwhile, around the Los Angeles metropolitan area, fires were started in Compton, Westwood, Beverly Hills, Hollywood, Koreatown, and the San Fernando Valley. Strangely, however, Simi Valley, scene of the trial, was quiet.

Throughout sections of South-Central, helmeted troops dressed in camouflaged combat fatigues stood watch at strategic corners, M-16s at the ready. The smell of charred wood permeated the air and everywhere lay the burnt and water-soaked remains left behind by looters. Melted video cassettes, shoe polish, plastic toys, torn magazines, singed boxes of disposable diapers, and cartons filled with soggy sanitary napkins were heaped outside a drugstore. From inside the blackened hulks of buildings came the sickening essence of spilled perfume, shaving lotion, toothpaste, mouthwash, hair spray, motor oil, and perhaps a hundred other substances ("Understanding the Riots," 1992).

Resulting Damage and Public Reactions

It would be weeks before a full assessment of the tragedy could be made. When the assessment was completed, Los Angeles would once again emerge as the site of the worst urban riot in history, exceeding in deaths and destruction the tragedies in its own Watts area in 1965 and in Detroit in 1967. There were 58 deaths and almost 2,300 injuries. More than 1,150 structures were destroyed. Approximately 10,000 small businesses were damaged. In most cases, rioters seemed to make little distinction between stores owned by whites, blacks, Hispanics, or Koreans. Whole city blocks were destroyed, and some residents had to walk many blocks to find stores still in business.

The majority of the 58 dead were between the ages of 18 and 50. All but one was a male. Twenty-seven were black, 17 were Hispanic, 11 were white, and 2 were Asian (1 was of unknown ethnicity). Of the 9,400 people arrested, about one half were Hispanic, one third were black, and 11% were white (the remainder, 2%, were "other") ("The Toll From the Riot," 1992).

The 1992 disorders became not only the most recent but, sadly, the worst mob violence to occur in the nation's poor urban neighborhoods in this century. If few people were surprised by the collective expression of anger over the King decision, many were shocked at the extent of the death and destruction. For liberals old enough to remember America's worst period of extended rioting in the mid- and late 1960s, the 1992 disorders confirmed their arguments that, in the years since the War on Poverty, poverty was winning. Conservatives, on the other hand, were reassured that billions of dollars of urban and social welfare spending by the federal treasury had failed to prevent the worst excesses of mayhem and disorder from reoccurring. Conservative Edward C. Banfield (1974) had written almost 20 years earlier that "there is likely to be more rioting for many years to come, and this no matter what is done to prevent it." During the 1970s and 1980s, Banfield predicted, there would be "frequent rampage-forays and some major riots." No matter how much effort we put into ending social injustices and racial discrimination, he argued, none of it will have "an appreciable effect on the amount of rioting" over the next 10 or 20 years (pp. 231-232). As it turned out, more money and more effort were expended on behalf of these ends. Urban rioting did indeed diminish over the next two decades. But sadly, it did not—and has not—ceased altogether.

However familiar the South-Central tragedy appeared to those who recalled the episodes of rioting during the 1960s, at least three significant differences were evident: First, never before in history had events precipitating urban interracial mob violence been visually recorded; until the 1992 Los Angeles riots police violence against an unarmed minority had never been portrayed to so many witnesses in such stark and graphic detail. Second, never before had large-scale rioting erupted in response not to a near-term event but to one occurring a full year earlier. As if to reinforce the effect of the video recording, the beating of white truck driver Reginald Denny by four blacks during the rioting produced another chilling tele-event. That strip was replayed endlessly during the trial of Denny's assailants in 1993 and conveyed a peculiar kind of tit-for-tat counterpoint to the theme of interracial cruelty.

A third variation from riots of the past appears to be in the racial and ethnic composition of participants and victims. In the past, the large majority of those arrested and the observed participants in looting, arson, vandalization, and other crimes were African Americans. Most of the businesses damaged, destroyed, or looted were white owned or white operated. Because of the large influx of Latin American immigrants into South-Central over the past two decades, those participating in the 1992 violence were more likely to be Latinos, with blacks still a major presence. Landlords and merchants of damaged properties were more likely to be Asian or Latino, although black businesses suffered significantly as well.

One year after the Los Angeles riots, a federal jury convicted two of the four police officers—Koon and Powell—of violating Rodney King's civil rights through use of excessive force (Rezendes, 1993). Although the other two defendants were found not guilty, the streets of Los Angeles remained calm that day. Within 2 years of the riot, however, it was clear that the massive recovery effort in South-Central had fallen short of almost everyone's hopes. Rebuild L.A., an organization that brought together business representatives, local government officials, and community leaders, had spread itself too thin in an attempt to please a multiplicity of interests. Trying to be all things to all people, the organization's 94-member board spent much of its time bickering about racial, ethnic, and partisan issues. Five people cochaired the board, including a white male, a black male, a Hispanic male, and an Asian female; leadership was thus diffuse. The group's mission was so multifaceted, it could not meet all or even most expectations. In its defense, it did assist in building new businesses and an employment training center in South-Central. By May 1994, it had succeeded in committing more than half of the $500 million in pledges it had received from corporations (Sims, 1994).

With time, Rebuild L.A. succeeded in limiting its mission to small business development, thus focusing on enhancing employment, personal income, and taxable real property. Ironically, by the time Rebuild L.A.'s hard-won lessons were in hand, federal officials in Washington had already committed the nation to a new program that would attempt to create similar organizations in other communities. Like Rebuild L.A., the federal initiative would have cities center action on a poor section; involve public, private, and community interests in devising responsive programs; and finance efforts from a combination of federal, state, and local government sources as well as business contributions and investments. A strong emphasis on private sector involvement and volunteerism in Rebuild L.A. is mirrored in the new legislation, the federal Empowerment Zones and Enterprise Communities program. Whether such efforts in other cities—absent forceful oversight from the federal government—will succumb to the indecisiveness of Rebuild L.A. remains to be seen.

Race and Rioting in America

No matter how devastating their consequences, outbursts of the type occurring in Los Angeles have been infrequent during the 20th century. When they have happened, however, many have been clustered over relatively short time spans. There has been some tendency for these tragedies to coincide with United States involvement in foreign wars. For example, major rioting broke out during World War I in East St. Louis, Illinois (1917), Chicago (1919), and Washington, D.C. (1919) shortly after the Armistice (Meier & Rudwick, 1969; Wade, 1971; Wallace, 1978). In East St. Louis and Chicago alone, at least 24 whites and 62 blacks died (Drake & Cayton, 1962; Janowitz, 1968; *Report of the National Advisory Commission on Civil Disorders,* 1968).

During World War II, major riots erupted in Detroit (1943) and in New York's Harlem (1943). A total of more than 1,200 injuries and 40 deaths were recorded in the two disasters. Hundreds of properties were damaged or destroyed (Marx, 1971; Sitkoff, 1978; Wade, 1971). On top of these events, the "zoot suit riots" victimized Los Angeles in 1943. Carried out largely by American servicemen who assaulted Chicano youth, some of whom wore flashy suits, the attacks were widely viewed as racially and ethnically motivated (Jones, 1969).

The worst period of rioting took place during the Vietnam War in the mid- and late 1960s.[1] Although major episodes of violence exploded in Harlem (1964), the Watts section of Los Angeles (1965), Chicago and Cleveland (1966), Newark and Detroit (1967), and Washington, D.C. (1968), hundreds of smaller tragedies also marred the era (*Report of the National Advisory Commission on Civil Disorders,* 1968). Yet urban mob violence has appeared even in peacetime, for example, in Harlem in 1935 (Wallace, 1978); in Miami in 1980, 1982, 1989; and, of course, in Los Angeles in 1992. If the historical context within which these tragedies occurred is not easily generalized, neither have the dynamics of the violence itself been consistent. Riots during and immediately after World War I and during World War II were invariably initiated by whites against African Americans, usually in areas such as downtowns or public parks. Whites often pursued their victims into minority neighborhoods, with beatings, shootings, and damage to homes and other properties the result (Brown, 1969; Drake & Cayton, 1962).

From the 1960s onward, however, riots tended to involve primarily
African Americans attacking properties (mainly business premises) and
sometimes civil authorities (mainly police) in or near predominantly black
neighborhoods. Although some observers predate this trend to the 1935
and 1943 Harlem riots, few disagree that the new pattern was established
by the 1960s (Janowitz, 1968; Meier & Rudwick, 1971; Wallace, 1978).
Interpersonal, interracial violence, with whites initiating action against
minorities, characterized earlier mob violence. Many of their victims had
lived in the city only a short while, having migrated from the South in
search of jobs. Urban whites, unused to African Americans and unwilling
to compete with them for jobs and housing, struck out with relatively minor
provocation. Later, when their presence in American cities was well
established, African Americans became more assertive during riots, attacking white symbols of authority and power such as the police and neighborhood shops. These events usually followed incidents in which African
Americans were arrested by police for minor infractions, followed by
actual or alleged police brutality toward the suspect. In some cases,
incidents of brutality by African Americans against nonblack bystanders
or passersby occurred.

These characterizations offer some sense of the gradual redirection of
interracial mob violence over the 20th century. But they leave unanswered
the questions of who riots and why. Are violent crowd reactions to individual episodes of racial conflict such as police arrests all that underlie the
death and destruction resulting from urban mob violence? Or are these
tragedies explained by more fundamental problems rooted in deprivation,
discrimination, and hopelessness? Adequate accounting of the substantial
literature on these issues is elusive. But the dominant parameters of debate
are illustrated in the work of a few key authors.

Most controversial among them is Banfield (1974), who has argued
that some African Americans riot in response to a youthful quest for
excitement or a "foray for pillage." Other African Americans do so in
righteous indignation as a spontaneous response to a perceived injustice.
Still others, Banfield argues, riot to advance a political principle or ideol-
ogy or to contribute to the maintenance of an organization. These are likely
to be planned rather than spontaneous outbursts. Obviously, excitement
and pillage imply self-serving motives for rioting, whereas reacting to an
injustice and advancing political ideals connote acts in pursuit of higher

ideals tending toward selflessness. Examining the Harlem (1964), Los Angeles (1965), and Detroit (1967) civil disturbances, Banfield concludes that most rioters were propelled primarily by excitement and pillage.

Fogelson, who examines the 1965 Watts tragedy in Los Angeles, reaches conclusions that do not support Banfield's quest-for-excitement or foray-for-pillage hypotheses. Instead, Fogelson (1969c) found that incidents of violence were legitimate protests against real grievances. Sears and McConahay (1973) examine several theories about riot causes and acknowledge that some who participate in the violence do so for reasons that are less noble than political protest. These authors conclude that most people become involved in riots as a result of their attitudes about politics and society. Like Fogelson, Sears and McConahay conceive of most of the rioters as motivated by higher ideals.

The *Report of the National Advisory Commission on Civil Disorders* (1968) concludes that urban mob violence in the 1960s grew most directly out of reactions to white racism. Among the effects of racism driving the violence, the report notes, were racial discrimination and enforced segregation, white flight from the cities to the suburbs, and the concentration of poor blacks in rundown ghettos.

Underlying many explanations for the causes of rioting then, is the view of beatings, lootings, killings, and property damage, however unfortunate, as a logical outgrowth of deprivation. Deprived of material needs and equal treatment, African Americans in particular are viewed as exercising conscious or unconscious political views. Although self-serving motivations may reside on the surface of unlawful acts, at their core is a sincere and selfless desire to express anger and outrage at the injustice of deprivation, according to many liberals and academics. Explanations for violence such as Banfield's (1974), which emphasizes self-serving motives, appear not to have been as widely subscribed to in contemporary government reports or in the scholarly literature. Yet there is little doubt of their popu- larity among the general American public (especially among non-African Americans).

A major element of deprivation theses is poverty, as measured by income, unemployment, poor housing conditions, and the like. Recent evidence, however, casts some doubt on the poverty explanation. Examining data from the 200 largest metropolitan areas (as well as data from riots in other nations), a study found little relationship between the outbreak of urban mob violence, on the one hand, and poverty or unemployment levels in the places in which these incidents occurred, on the other (L.A. and the Economics of Urban Unrest, 1994). Yet when riots did break out, their intensity was likely to increase as the incidence of poverty or unemployment increased. This research shows that stable communities (i.e., those with higher shares of longer-term residents) are less likely to riot than less stable ones. Moreover, communities with higher rates of home ownership and self-employment (i.e., where residents have a greater stake in the area) are also less likely to riot. This study suggests that public policy might more effectively concentrate on increasing stability, home ownership, and self-employment than on reducing poverty, per se, in attempts to diminish the likelihood of urban mob violence. A single study cannot by itself confirm a new hypothesis, however, much less overturn an old one. Furthermore, some might convincingly argue that data from the 1960s pose an insufficient basis on which to frame 1990s policies. Nevertheless, the study merits attention and is provocative in its implications.

This book cannot and does not attempt to contribute to, much less resolve, the dialogue on the causes of urban interracial mob violence. Rather, it takes as a point of departure that at least some of the conditions associated with deprivation primarily among African Americans—especially the various dimensions of poverty—are essential to any explanation of why people engage in riot behavior. Although poverty is accepted as necessary to such an explanation, it is not sufficient.

More central to the purpose of this book is not whether a consensus exists on the causes of urban mob violence, but that many who have been influential in federal policy-making processes believe that poverty is a central element (or claim to believe so). As the chapters ahead will show, the poverty and deprivation theses have persisted in their ability to excite controversy on Capitol Hill for at least a quarter century. But the strength of those ideas has been a dominant force in shaping public policy reactions to outbursts of mob violence both in the 1960s and in the 1990s.

Note

1. Some argue that ghetto rioting persisted well into the 1970s (Feagin & Hahn, 1973).

3

Civil Rights and Uncivil Riots, 1964 Through 1966

In the period between the 1943 riots and those that arose in the 1960s, virtually every large city gained a significant percentage of minority residents. An entire generation of white urbanites had grown up, if not in close proximity to black families, at least with the assumption that a multiracial urban society was here to stay. Consequently, although racism had hardly disappeared, whites were generally less intolerant of, and in some cases more sympathetic to, the black condition in America (Meier & Rudwick, 1971). If white Northerners had grown up with little or no interaction with blacks, the reason was assumed to be primarily economic in nature. Even though the ravages of discrimination—official and unofficial—had not been eliminated, differences in access to education, adequate housing, and jobs had exacerbated socioeconomic conditions between the races. These disparities were translated into densely populated, predominantly black central city neighborhoods and rapidly spreading, almost entirely white suburbs. City halls and courthouses were still dominated by white officials. Public school systems and fire and police departments were

14

overwhelmingly staffed by white workers. Yet two entirely new ingredients had been added: a president who was about to seek historically unprecedented reforms in the welfare of minorities and poor people and a charismatic and gifted man of God who was poised to become a legend.

The Freedom Train of Civil Rights

If anyone predicted the nation's most extensive period of rioting at the beginning of 1963, their warnings went unheard—or, perhaps, unheeded. For many Northerners, interracial mob violence was something that happened in the South. Although serious racial disorders occurred that year in Chicago and Philadelphia, these were modest in their scope and easy to dismiss as simple aberrations. But in many southern states, the civil rights movement was in full swing. Under the guidance of organizations such as the Southern Christian Leadership Conference (SCLC), the National Association for the Advancement of Colored People (NAACP), the Congress of Racial Equality (CORE), and the Student Nonviolent Coordinating Committee (SNCC), efforts were under way to register blacks to vote. Sit-ins, boycotts, and picket lines were employed by blacks and sympathetic whites to voice anger at segregated lunch counters or municipal buses and at white-only public facilities such as swimming pools and rest rooms. Cities such as Birmingham, Savannah, and Cambridge (Maryland) experienced violence between whites and blacks. Civil rights marchers were sometimes met by local whites, some of whom shouted curses at demonstrators and punched, shoved, and kicked the protesters. Frequently, white police were sympathetic to the troublemakers (Fairclough, 1987). Among the worst incidences of interracial violence in the South was the bombing of a black church in Birmingham in 1963, killing four black girls. The assassination of black civil rights worker Medgar Evers followed in 1963. The following year, three civil rights workers—two white and one black—were murdered by white vigilantes, their bodies buried in Philadelphia, Mississippi. Other incidents of racial violence were scattered across the South (Sitkoff, 1981).

These events reinforced a growing restiveness in northern cities. Yet northern blacks had the same general freedom to vote as whites. De jure segregation had faded. Blacks and whites used the same rest rooms, eating facilities, buses, and other public spaces. Mayors, governors, and others found it hard to compare race relations in their cities with those in the deep South. Nonetheless, blacks knew only too well that public service jobs as teachers, firefighters, police officers, ambulance drivers, transit workers, and city hall employees were largely controlled by white-dominated unions and white-controlled political organizations. Few, if any, blacks appeared in corporate boardrooms or executive suites. Catholic and Protestant churches—outside the traditional black sects—were run by white clergy. Downtown retail stores rarely hired blacks.

Moreover, racially segregated housing was widespread. Through outright discrimination or a panoply of ruses and diversions, many white landlords were able to prevent blacks from moving into apartments and houses. Many real estate agents steered whites to housing for sale in white neighborhoods and blacks to lesser units in black neighborhoods. Others were known to practice scare tactics such as block busting. Some real estate agents convinced home sellers that blacks would depress property values and encouraged them to sell at depressed prices. In some cases, these agents bought the properties themselves, selling them to black buyers at inflated prices. Increasingly, public and parochial schools reflected the racial composition of the neighborhood in which they were located. Most inner-city school enrollments became overwhelmingly composed of whites or blacks; teachers and administrators in nearly all schools were white.

By 1964, it was apparent to vast numbers of blacks and some white liberals that although racial conditions in the South were dreadful, those in the North, although less blatant, posed many of the same day-to-day constraints on black freedom and opportunity. Just the year before, Martin Luther King, Jr., had delivered his renowned "I Have A Dream" speech at the Lincoln Memorial in Washington DC, before a crowd estimated at more than 200,000 and millions of television viewers (Sobel, 1967).[1] Yet few in King's audience had any idea that the 20th century was about to witness some of the most monumental civil rights advances since the end of the Civil War.

In mid-summer 1964, President Lyndon B. Johnson was busy campaigning for the November presidential election. Less than a year earlier, President John F. Kennedy had been assassinated in Dallas; as vice president, Johnson had succeeded him. Without an electoral mandate behind him and overshadowed in the public mind by Kennedy's popularity and charisma, Johnson nonetheless sought support among liberals and working class white and black constituents who had voted for his predecessor in 1960. Facing conservative Senator Barry Goldwater, who was awarded the Republican Party presidential nomination on July 15, Johnson was busy crafting a series of proposals for new legislation to address civil rights and poverty in America. Some of these ideas had surfaced in the New Frontier years, but others were forming in response to the growing influence of the civil rights movement. The nation's cities had been relatively quiet and little of the sense of urgency that was to arise later was yet evident. Urban interracial mob violence in Detroit and Harlem had occurred two decades earlier and was now a distant memory for most people. Although the early 1960s had brought some crowd violence associated with civil rights activities to some southern cities, few northerners associated these outbursts with conditions in their own cities (Kearns, 1976; White, 1965).

Harlem, 1964

On July 16, 1964, a black teenager in Harlem was shot and killed by a white police officer. The officer maintained that he had been defending himself against a 15 year-old who attacked him with a knife. Soon, windows were broken by black teenagers who gathered until a large force of officers dispelled the crowd. Soon after, a protest march organized by CORE against the killing of three civil rights workers in Philadelphia, Mississippi, deteriorated into another struggle with police. When this incident was over, 1 person had been killed and 12 police officers and 19 citizens were injured. Thereafter, continual flare-ups resulted, not only in Harlem but also in the Bedford Stuyvesant section of Brooklyn. Objects were thrown at police and firefighters, police retaliated with gunfire, passersby were beaten, and shops were looted. Between July 18 and July 23, 1964, 1 person was killed and 144 were injured (Fogelson, 1969b; White, 1965). Within a week, four more people would die in Rochester, New York, where similar violence left hundreds injured, almost 1,000 arrested, and nearly $3 million in property damage (Button, 1978). Disorders followed in Jersey City, Paterson, and Elizabeth, New Jersey. The summer of 1964 closed with a 2-day riot in Philadelphia that grew out of a police attempt to move a stalled car out of a busy intersection in a black neighborhood (*Report of the National Advisory Commission on Civil Disorders,* 1968). When the violence subsided, mayors, council members, governors, legislators, and members of Congress could only hope that these were isolated incidents and not harbingers of future patterns.

Scarcely 5 weeks after the Harlem discord, Congress enacted legislation establishing the Office of Economic Opportunity (OEO) in the White House (Frieden & Kaplan, 1975). A host of OEO social welfare programs would follow under an $800 million congressional appropriation (Haar, 1975). Yet neither Congress nor the White House had intended the office to be an exclusively urban-centered initiative. Thus, the public image of the OEO vacillated between urban and rural identities. Only later in the decade would the OEO become widely identified with Johnson's efforts to improve life for urban blacks and other minorities.

Johnson Secures a Mandate

In November, Lyndon Johnson's landslide defeat of Barry Goldwater gave him the mandate he had sought. Moreover, the accompanying congressional elections had put more liberal northern urban Democrats in Congress than at any time since the New Deal (Mollenkopf, 1983). Over the next year, plans for Johnson's Great Society congealed and new legislation was enacted in areas such as education, housing, health care, crime control, and juvenile delinquency.

But even as signs of racial and social progress were beginning to appear, it seemed that the nation could not move fast enough to head off the growing rancor in its cities. Adding to tensions in the cities were incidents such as the assassination in New York of Malcolm X in February 1965. The event did not lead to mob violence, perhaps only because the victim's killer proved to be another black man (Sobel, 1967). The civil rights march from Selma to Montgomery, Alabama, followed shortly thereafter; millions of television viewers were shocked by the sight of police dogs and cattle prods being used to assault peaceful marchers, both black and white (Fager, 1974).

Watts, 1965

In August 1965, the ink was barely dry on the newly signed federal Voting Rights Act when the nation experienced its worst outbreak of urban interracial mob violence. Enfranchising perhaps a million black voters throughout the nation (Haar, 1975), the law had little effect on the crowds in Los Angeles that gathered at the site of a routine arrest by police on August 11. Stopped for driving while intoxicated, a young black man was taken into police custody. Soon rocks and other objects were being hurled at passing cars. White drivers passing by were pulled from their cars and beaten. Automobiles were upended and burned (Cohen & Murphy, 1966).

Similar events occurred on the following evening; the next day, crowds began smashing windows and looting and burning stores. Soon the violence spread to the Watts area 2 miles away. Hundreds, perhaps thousands, took part in looting stores there. Firebombing with Molotov cocktails was extensive. The National Guard was called; 30 hours after its arrival, order was finally restored. The military units and local police deployed firearms extensively in response to numerous incidents of sniper fire from rooftops and windows in Watts. When the lawlessness was finally over on August 16, the tally of violence was 34 killed and 1,032 injured (Fogelson, 1968). Nearly 4,000 people had been arrested and $35 to $40 million in damages were sustained. More than 600 buildings were damaged. The Watts disorders had surpassed even the 1943 Detroit riot for sheer death and destruction (Cohen & Murphy, 1966; *Report of the National Advisory Commission on Civil Disorders,* 1968). Although there were fewer arrests due to riots in 1965 than in the previous year, the disastrous Watts episode contributed to the largest cumulative number of deaths (43 persons) and woundings (1,206 persons) occurring nationally in a single year due to such events (Wikstrom, 1974).

By the time Los Angeles cooled down, many in Washington and in city halls throughout urban America were growing increasingly uneasy that the looting, torching, and destruction were not simply aberrations. Racial anger that at first appeared to be confined to southern communities undergoing desegregation now threatened to emerge in a fearsome new form of destruction in northern cities. And even though the violence in Harlem, Rochester, Los Angeles, and other cities erupted under ostensibly different circumstances than the beatings, hosings, and cattle proddings administered by small town sheriffs and working class whites, it was another sign that the fragile threads of racial accord in the nation were becoming unraveled

Americans viewed the rising tide of discord in the cities from different frames of reference. Older generations could recall the spate of disorders involving minorities in Detroit, Los Angeles, and Harlem during World War II. For them, the newer disturbances were all too familiar. But these earlier events all occurred within a few months of one another and by no means constituted a trend. For a younger generation of Americans, however, there was nothing with which to compare the Watts violence. The baby boom generation, born in the years after the war, had never encountered such intense racial animosity nor such widespread destruction. For most whites, regardless of age, there was confusion over the causes of these tragedies. In the face of monumental progress in civil rights over the previous decade and a half, whites were troubled by the television images of northern blacks resorting to violence, just as some southern whites had been doing, to achieve their ends. With segregated schools outlawed, the Voting Rights Act in place, and the OEO developing new programs, surely conditions were improving. Hundreds of millions of dollars had already been spent to tear down slum buildings, to relocate families and businesses to (presumably) better housing, and to build new highways, homes, shops, offices, and public facilities. From the perspective of many whites, these were signs that the cup was at least half full. For too many black citizens, though, the cup was at least half empty.

Some whites reacted with resentment and saw the riots as confirmation of the putative violent, lawless nature of black people. Others were troubled by the foundation of grievances to which black people pointed in explaining why the melees had occurred. Many older black people deplored those events and feared that they would be met by a white backlash, thus reversing gains already made. Younger nonviolent blacks were more likely to understand the source of the rage and frustration among their rioting peers. Discordant though the voices of black and white leaders were on these issues, an even more troubling refrain was being uttered by a small group of black radicals such as Stokely Carmichael, H. Rap Brown, Bobby Seale, and others who championed more drastic action. The Black Panther party, although always encompassing a tiny share of the black population in America, grew in popularity among many younger urban blacks (Anthony, 1970; Marine, 1969).

In the White House, President Johnson and his inner circle agonized over the rising signs of urban unrest. The president felt betrayed by the uprising in Watts and soon became convinced that the riots were spurred by outside agitators traveling from city to city sowing seeds of distrust (Kearns, 1976). In a speech on August 20, 1965, Johnson compared the rioters in Watts to Ku Klux Klan members. Both, he insisted, were "law-breakers, destroyers of constitutional rights and liberties, and ultimately destroyers of a free America" (quoted in Califano, 1991, p. 63). In the years ahead, the president's bitterness would only increase.

A Critical Turning Point

By October 1965, President Johnson had set up a task force to make recommendations on organization and programs for the newly established Department of Housing and Urban Development (HUD). Not to be outdone by his White House predecessor, who appointed several academics to positions in his administration, Johnson chose as chair of his task force a young political scientist from the Massachusetts Institute of Technology, Robert C. Wood, who would soon become second in command under Secretary Robert Weaver at HUD. Wood had served Johnson ably as chair of the Task Force on Urban Problems and the president trusted him.[2] An immediate charge from the White House to the new task force was to examine a proposal in May 1965 by United Auto Workers Union President Walter Reuther to President Johnson (Haar, 1975). Reuther proposed a six-city federal demonstration program that would promote cooperative planning and execution of self-designed programs to revitalize the poorest areas of each city (Haar, 1975). Rather than a one-size-fits-all policy approach from Washington, this approach would challenge each partici-pating city, through coalitions of diverse interests, to design individualized, tailor-made programs to address urban problems in the participating neigh-borhoods.

Wood's task force carried on much of its work in secret, squirreled away in downtown Washington in bureaucratic offices (Haar, 1975). Ever mindful of rising urban discord, the group "was shaken by the Watts violence" (Button, 1978, p. 65). The sheer extent of the damage and the growing sense that urban rioting was becoming a trend helped clarify the crisis nature of the task force's mission (Sundquist & Davis, 1969). It was becoming increasingly obvious that something out of the ordinary would have to be done. Clearly, those who were disfranchised, discriminated against, and angry should be among the beneficiaries of any new HUD initiative. As the task force staff labored that fall, a mood of foreboding about the future was brewing in Washington.

The task force report prescribed a bold and remarkable congeries of actions to be considered by the president. At its heart was a strong emphasis on renewing the physical fabric of what were then commonly called urban *slums,* a term out of fashion today. Yet there was little mention of revitalizing shops, factories, offices, municipal facilities, and the like. Instead, the report focused on the need for new and rehabilitated housing. Issues such as transportation, air and water pollution, social service needs, federal programmatic coordination, racial and social discrimination, and the like were viewed as attendant but not central. Housing and neighborhood renewal were clearly the centerpiece in the task force's view. The task force's goal not only recognized the dire need for improved dwelling opportunities among the urban poor but it was also a direct response to one of the chief complaints about the federal Urban Renewal program. Critics charged that most local redevelopment authorities concentrated their activities in downtowns and industrial areas rather than in neighborhoods. Authorities built offices, hotels, shops, and other commercial and industrial facilities that promised to generate jobs and to yield higher tax revenues than housing. There could be no mistaking the priorities of the task force members, however; theirs was a federal commitment to the home, the family, and the neighborhood.

The seeds of Reuther's original proposal were sown in the task force's report. It proposed an experiment involving 62 cities of varying sizes selected on a competitive basis from applicant communities throughout the nation (Haar, 1975). Each participant would receive federal funds to prepare a plan primarily for revitalizing housing for poor families and secondarily for providing related shopping, employment, public facilities, and other opportunities. When plans were approved by the federal government, money to carry them out would be made available, both from the existing array of federal programs and from a newly created subvention called "supplemental funds." Supplemental funds would be used to finance total project costs in excess of those covered by categorical grants under existing federal programs. For every dollar of financing not provided by the categorical programs, supplemental funding would contribute 80 cents and the appropriate state and local governments would contribute 20 cents. Nothing like supplemental funding had ever been attempted by Washington before. It was intended to supply cities with additional resources for urban development. But it was also intended to be a financing mechanism much more flexible and discretionary than conventional categorical grants, which most municipal officials felt were overly restrictive in their limits on how funds could be spent.

The report proposed a 5-year demonstration program with enhanced citizen participation, attention to good design and historic preservation, open housing and desegregated neighborhoods, better public transportation options, and provision of supportive social services along with physical renewal. It urged attention to local property taxes, building codes, and building trade practices that unnecessarily heightened the cost of inner-city reinvestment.

Although the report focused on the problems of inner cities, it also proposed that six to ten additional demonstrations be funded to test new methods of encouraging more orderly metropolitan growth, including planned communities, regional transportation planning, and interlocal coordination of development decisions. It also recommended reforms to the existing federal categorical grant system through block grants to assist in paying for a broad swath of municipal operating expenses not covered by existing programs.

In retrospect, what is most remarkable about the 1965 task force report is how many of its innovative recommendations would later be implemented in identical or similar forms. In fact, the report would prove to be highly prophetic of federal policy development, under both Democratic and Republican presidents, for at least another decade. In part, of course, this was due to the wide net cast by the task force among participating scholars, analysts, and other experts on urban affairs. But the relatively high rate of success with which so many of these proposals later became law was due also to the sharp analytical eye the task force and its staff and consultants were able to apply to the complicated urban conditions of the day. Whatever the potential efficacy of its many recommended actions, the report could hardly be faulted for its insights on the nature of underlying weaknesses among existing federal policies. The 1965 task force report would prove to be a crucible of sharp thinking, containing some of the most innovative policy recommendations of the day.

There was one conspicuous omission in the report, however. Considering that poor neighborhoods in New York, Rochester, and Philadelphia had experienced major civil disorders only the year before—and Los Angeles had exploded only months earlier—the report was entirely silent on these events. There is no mention of the destruction anywhere in the document. Doubtless, there was a political rationale for this: The task force, and perhaps the president and his staff, did not want to appear to legitimate the lawlessness associated with the riots. Critics could too easily seize on the report's recommendations as evidence that rioting pays. Perhaps, too, the White House sought to avoid bringing undue attention to the violence, thereby possibly alarming the press. Yet there is no doubt that the task force report was intended in part as a response to urban interracial mob violence.

Curiously, the task force report made almost no mention of what would later be called the *demonstration* or *model* neighborhood. There is only a single reference to a "designated area" and another to "the total environment of the area affected." Otherwise it is not at all clear that the task force's primary proposal was to be concentrated on a single poor neighborhood in the communities selected for participation. Instead, there is a vagueness to this issue, leaving plenty of room for the White House and Congress to design a program that might target one, two, or more blighted sections of a city for action. Yet the report called for the "concentration of available and special resources in sufficient magnitude to demonstrate swiftly what qualified urban communities can do and can become" (Haar, 1975, p. 295).

The Centerpiece:
Demonstration Cities

Most of the task force's recommendations became the basis for a new measure President Johnson introduced 6 weeks later in his budget message to Congress (Semple, 1966i). The president's Demonstration Cities program, as Model Cities was first called, would avoid the liabilities of the federal Urban Renewal program, Washington's chief slum fighter. The new venture would not involve massive relocation of households and businesses, would not demolish large areas of poor neighborhoods, and would not concentrate only on physical renewal. It would do the following:

- Provide assistance to cities to devise their own programs (rather than those handed down from Washington with rigid procedural guidelines and regulations) to bring "physical rehabilitation of slum areas" and "social rehabilitation of the people in them."
- Focus this assistance on a single neighborhood in each participating city in which poverty and minority needs were high.
- Ensure the involvement of residents and businesses in the target area in the planning and execution of their demonstration program.
- Last for 6 years. Afterward, Congress would examine each city's performance in the program and devise long-term federal initiatives from the lessons learned. These new programs would presumably be open to all U.S. cities.

The Demonstration Cities legislation proposed spending $12 million to allow selected cities to prepare plans for their target neighborhood. After plans were approved by HUD staff, $2.3 billion would be allocated over 6 years to translate the plans into action. Cities would compete for a planning grant; 60 to 70 communities would be selected. The Demonstration Cities plans could involve housing renovation or construction, job training, innovative educational programs, day care services, youth recreation, neighborhood beautification, and an almost limitless list of other goods and services. The message here was that Washington would trust citizens and city halls to make better decisions about needed programs and delivery systems than federal bureaucrats could devise through complicated program guides. President Johnson and his aides were only too aware that the growing number of civil rights and citizen participation groups in American cities chafed under the seemingly heavy hand of the federal agencies and an impenetrable web of guidelines and reporting requirements mandated under categorical grants. They were also aware that low-income and minority residents often disagreed with city hall and federal priorities.

One way to derail this disaffection was to turn over more responsibility for program planning and execution to citizens while ensuring that city halls—dominated mostly by Democrats—remained in control. President Johnson did not want to repeat what he came to view as the excesses of Head Start, Job Corps, the Community Action Program, and other OEO activities. These efforts irritated many city officials who believed that OEO gave too much control over resources to nonelected people.

In addition, President Johnson realized that Congress would never appropriate enough money, in addition to funds already being spent under Washington's many categorical programs, to pay for all the work needed in the nation's urban slums. From the start, then, the Wood task force had proceeded under the assumption that the president would look for ways to involve existing programs such as Urban Renewal, parks and recreation, sewer and water facilities, and employment training in the revitalization strategies in each Demonstration Cities neighborhood. Therefore, each approved application would incorporate proposals funded by conventional federal grants as well as the new supplemental grants. Herculean efforts by HUD would be required to convince other offices and departments such as OEO, labor, commerce, and health, education, and welfare to cooperate in the coordination effort. As presented, the legislation would allow a city to use, for example, a Department of Labor categorical grant for employment training to pay up to 75% of program costs; four fifths of the remaining 25% of costs could be covered with funds from a Demonstration Cities supplemental funds grant. In effect, up to 95% of program costs would be covered by federal money, leaving state and local treasuries to make up the rest.

From Fanfare to Fading Flower
in the 89th Congress

The next day, the bill was introduced into the House by Representative Wright Patman (D-TX). Paul Douglas, an Illinois Democrat, was the lead cosponsor in the Senate. A week later, mayors from 25 major cities were briefed by HUD Secretary Robert Weaver about the proposed program. Concerns were expressed by New York Republican Mayor John Lindsay— then one of the most prominent U.S. urban officials—that too little money would be spread over too many cities. Over the 6-year life of the demonstration, only about $40 million would reach each participating city, he warned. But some mayors liked the mixing of social and physical revitalization efforts and the focus of the proposal on the neediest sections of the city (Semple, 1966g). On February 21, Ada Louise Huxtable gave the proposal rave reviews in her *New York Times* column. Demonstration Cities was already "acclaimed by planning experts as one of the most comprehensive and far seeing Government programs produced in this century to deal with the urban crisis" (Huxtable, 1966, p. 12).

The first watershed for Demonstration Cities was hearings before the Housing Subcommittee of the House Banking and Currency Committee from February 28 to March 25 (Haar, 1975). Within a few days, Mayor Lindsay publicly endorsed the proposal with two caveats. First, proposed funding was so low, he argued, that his own city could easily consume the entire amount slated for planning and carrying out the national effort. Second, Lindsay found it "distasteful" that he would be able to choose only one Demonstration Cities neighborhood from among at least six in New York that merited the program. Appearing at the same meeting with Lindsay, Detroit Mayor Jerome P. Cavenaugh, president of the National League of Cities/U.S. Conference of Mayors, concurred (Semple, 1966f). Within 16 months, Detroit would set a new and unenviable record for the worst carnage and destruction of any single urban riot in American history.

As it became clear to members of the House subcommittee that many mayors favored the intentions of the legislation, it also was evident that several held strong reservations. The issue of the proposed amount of congressional appropriations continued to dog the initiative. A second concern was the supplemental grant mechanism. Some subcommittee members feared that it was ripe for exploitation by participating cities. As written, the legislation would encourage cities to find ways to tie federal categorical grants into their Demonstration Cities plan even though the grants have little bearing on or relevance to the selected neighborhood. In making the connection, cities would increase the amount of supplemental funding for which they would become eligible. But the bill would also permit cities to count supplemental funds as part of the required matching local share for several categorical grant programs. Thus, the larger the supplemental fund allocation to each community, the less it would have to take from locally derived revenues to help pay for these other programs (Semple, 1966l). Several members of the House subcommittee were uncomfortable with the spending implications of the bill; by the time the hearings closed, support had cooled.

In mid-May, an editorial in the *New York Times* opined that the Demonstration Cities program was dead ("No Demonstration Cities," 1966). Aside from concerns already expressed in Congress, there was growing suspicion among cities that if the program was enacted, HUD bureaucrats would steer other agencies' categorical grant monies to Demonstration Cities neighborhoods, leaving mayors with little funding for projects elsewhere in their cities. This diversionary tactic, some believed, could become a classic example of robbing Peter to pay Paul. A corollary issue was the fact that the new measure would not spread federal largesse around the nation's congressional districts as generously as many conventional urban aid programs. There were to be only 60 to 70 winners; many in Congress wanted assurances that at least one city in their home district or state would be among the triumphant (Semple, 1966h).

Also bedeviling the Johnson bill was the fact that the new subcommittee chairman, William A. Barrett (D-PA), shied away from the kind of firm leadership that his predecessor, Albert Rains (D-AL), had exercised. Barrett's approach to subcommittee members was to seek consensus. Rains, who had left the House a year earlier, had exercised his authority and established clear goals at the beginning of hearings, followed later by compromise if and when the need arose. Subcommittee members thus soon received the impression from Barrett that there were few conciliatory limits beyond which he could not be pushed.

An Attempt to Revive

Reacting to the *New York Times* editorial ("No Demonstration Cities," 1966), President Johnson spoke publicly of the bill as " 'the most important domestic measure before the Congress and to the future of the American cities' " (quoted in Califano, 1991, p. 131). As if to confirm the bill's high priority, he assigned aides Larry O'Brien and Joseph Califano to shepherd the measure through Congress. In late May, HUD congressional liaison staff, headed by Sidney Spector, began a lobbying campaign to resurrect the bill. By then, the Senate Housing Subcommittee had completed its hearings on the measure. But with doubtful support in Congress, the House was not prepared to vote positively on the measure. Aided by Wood, now undersecretary at HUD, and HUD Deputy Assistant Secretary Charles Haar, a former member of the Wood task force, the HUD team churned out memos, correspondence, and phone calls on behalf of the measure, countering each point of opposition. A campaign to solicit endorsements from interest groups and federal agency heads followed.[3]

One reason why key members of Congress had been unwilling to put muscle behind Demonstration Cities was that at first the president himself had not appeared committed to it. After much fanfare in his budget message in January, he became preoccupied with the growing demands of the Vietnam commitment and other issues. Sensing an opportunity, Housing Subcommittee Chairman Barrett rewrote the bill, turning Demonstration Cities into a more conventional something-in-it-for-everyone program. Known generically as the "boodle bag" approach, Barrett's omnibus bill was frowned on by the White House. To convince Barrett to drop his amendment, Vice President Hubert Humphrey was pressed into service. He called the Pennsylvania congressman during a subcommittee meeting, reassured him that his efforts were appreciated, and insisted that the president felt that the measure had to survive substantially as proposed. Surprised that the White House was strongly behind the measure, Barrett returned to the subcommittee, "steered the conversation to some relatively innocuous amendments, and then adjourned the meeting until June 21" (Semple, 1966k, p. 1). With an additional 3 weeks of time now secured, Califano and O'Brien were poised to reinvigorate the campaign to secure passage of Demonstration Cities. They met with Barrett and argued for more than an hour urging him to advance the bill. Finally, he agreed to do so but wanted the Senate to pass it first (Califano, 1991). Thus, White House attention shifted to the other side of Congress.

The president decided to try to recruit Maine Democratic Senator Edmund Muskie to floor managerial duty.[4] Senator John Sparkman (D-AL), who normally shepherded urban legislation through the Senate, was preoccupied with a tough reelection campaign in his home state. Moreover, he did not want to give ammunition to opponents in his conservative home state by being associated with a liberal urban poverty measure. In any case, Muskie showed little enthusiasm for the task when first approached. Johnson then decided to appeal to the Maine senator's political instincts. A letter from Patrick Healy, head of the National League of

Cities (NLC) staff in Washington, urged Muskie to act as floor manager of the bill in the Senate. The NLC represented 13,000 municipalities in all 50 states and was widely regarded, along with its sister organization, the U.S. Conference of Mayors (USCM), as the reigning authority in the capital on urban interests.

Healy noted that the NLC leadership endorsed the basic principles of Demonstration Cities. He also assured the senator that the White House wanted the Demonstration Cities program to include cities of all sizes, not just large cities. This matter was of interest to Muskie; Portland, the largest city in Maine, Portland, had a 1960 population of less than 73,000 people. By this time, the bill was identified primarily as an effort to assist the poor and minorities in the largest, most riot-prone centers of the nation. Not only did Muskie's home state have no large cities, happily it had had no riots either. Maine's black population—2,800 in 1970—was less than 1% of the state's total population. If the measure were passed, Muskie's urban constituents in Portland, Lewiston-Auburn, Waterville, Bangor, and smaller cities were unlikely to be eligible for a Demonstration Cities grant. Healy's letter made it clear, however, that the Johnson administration was willing to expand eligibility for the measure to entice the Maine senator to lend his prestige to it (P. Healy, personal communication, undated; Healy, 1974[5]). Earlier, Califano had reminded the president that Muskie might not be interested in the task, given that Maine had no large cities likely to qualify for a Demonstration Cities grant.

> Well he has one now, replied Johnson.
> What one? asked Califano.
> Whatever one he wants, Johnson said, laughing. (Califano, 1991, p. 132)

(As it turned out, two Maine cities were awarded Model Cities designations in the demonstration program that eventually was enacted.)

This was not Muskie's only concern. He felt that the Demonstration Cities funding formula was too complicated. Wood, Haar, and Hilbert Fefferman, HUD Associate General Counsel, tried to clarify this issue (personal communication, June 9, 1966). They sought to reassure the senator that supplemental funds were intended to permit flexibility and innovation for localities in addressing particular social, physical, and economic problems in the Demonstration Cities neighborhood. They insisted that supplemental funds were not to be a congressional blank check with no strings attached. The three HUD executives acknowledged that suspicions existed in Congress that Demonstration Cities communities would seek to tie every federal categorical grant they received to the Demonstration Cities area so as to hike the size of the supplemental grant they would be eligible to receive. This memo suggests provisions to discourage such an outcome (personal communication, June 9, 1966).

In mid-June, the USCM approved a resolution at its annual meeting endorsing Demonstration Cities (Semple, 1966m). Robert Weaver, who spoke to the gathering, reassured the mayors that HUD would seek additional appropriations for Urban Renewal so that no existing funds would be earmarked for use in Demonstration Cities neighborhoods (Haar, 1975). Less than a week later, the USCM leadership informed Muskie that it had formally endorsed the Demonstration Cities legislation (personal communication, June 14, 1966).

With leaders of both the NLC and the USCM behind the measure, Muskie knew that the Demonstration Cities' survival hopes were enhanced. Still, the USCM cautioned Muskie not to allow Congress to divert resources from categorical programs such as Urban Renewal and the Community Action Program to this new initiative. It also opposed any attempt by nonsupporters to limit the program to providing planning monies only (H. Mields, personal communication, June 14, 1966).

Meanwhile, in an effort to persuade Muskie to accept sponsorship of the bill, Haar, Wood, and White House aides dispatched by Larry O'Brien and Joseph Califano met with the senator in mid-June (Semple, 1966k). But the meeting produced little progress and Muskie grew more irritated. The bill was needlessly complex, he insisted, and the funding formula was confusing. Muskie finally reached a decision; on June 16, he sent a letter to the president. He would not manage the Demonstration Cities bill if "it involves battling for the legislation as sent up by the Department of Housing and Urban Development." He listed four problems that would have to be resolved to gain his sponsorship:

- Republicans could attack the supplemental grant idea as a "back door" effort to extend block grants to cities. (Republicans had resisted the block grant because such grants would give taxpayer money to city halls with relatively few strings attached.)

- There had never been a record of cooperation between and among federal agencies. It seemed unlikely that HUD could change this pattern so that each successful city recipient of a Demonstration Cities supplemental grant could draw on a host of other federal grant programs to carry out its plan.
- The chances of enacting a strong urban bill were slim because the Democrats were vulnerable to Republican charges of big spending without "immediate and observable impact." Muskie noted that the measure, if passed, would have little chance of affecting "our crisis cities this summer." In effect, Muskie worried that Demonstration Cities would set high expectations but would not be able to contain further urban violence as the summer unfolded.
- To tone down the bill further, Muskie recommended reducing the number of grant recipients to "about 25 crisis cities this summer." This would make the program look a bit less like another federal pork barrel program and would allow HUD to put greater funds into fewer cities, enhancing the program's effect (E. Muskie, personal communication, June 16, 1966).

Responding to the Riots

Muskie had based his four points on his own insights and also on those of his office staff. An internal memo pointed out that Congress would not be able to enact the measure any sooner than mid- or late July. Therefore, the program could not help the nation's cities that summer. Yet, the memo notes, the administration expressed "two basic objectives" underlying the legislation:

- A "long-range program of metropolitan and urban planning with coordinated social and physical redevelopment of our communities."
- "Rapid action to avoid repetitions of the Watts riots in cities such as Washington, New York (Harlem), Chicago (where violence has already erupted), and Atlanta" (D. Nicoll, personal communication, June 15, 1966).

The president was not sanguine about the capacity of Demonstration Cities to stop riots. He likely wanted to be in a position to point out, in the likely event of further riots, that his administration had responded to the call of frustrated minorities and made an attempt to cool their furor. Johnson was growing ever more fearful of urban riots, even to the point of suspecting a nationally organized conspiracy. Two years later, the Kerner Commission report pointed out that it uncovered no evidence of central orchestration (*Report of the National Advisory Commission on Civil Dis-*

orders, 1968). Yet Nicolas Lemann (1991) cites an undated conversation Johnson had with Katherine Graham of the *Washington Post*. President Johnson told her that the Federal Bureau of Investigation (FBI) was aware of when and where each riot would occur. To that point, each riot had occurred just as the FBI had predicted. Whether Johnson's claims were based on fact is doubtful. That J. Edgar Hoover's FBI surveyed cities for riot potential is undeniable. That Johnson may have sought to plant his assertions in the *Post* through Graham is plausible. In any event, there is evidence that the White House was growing ever more traumatized by minority violence in American cities.

The effect of urban violence during the mid-1960s on the evolution of the Demonstration Cities program cannot be overemphasized. In particular, the Watts explosion in 1965 was "so vast, so awesome, so devastating, and so widely reported," says historian Fogelson (1968), "that henceforth there could be no doubt that a distinct pattern of summer violence was emerging in the black ghettos" (p. 329). Haar (1975) characterizes the national mood as apocalyptic. The nation, he says, "seemed beset by the puzzling violence of the summer of 1966" (p. 83). This state of affairs, Haar concludes, was the first condition giving rise to development of the Demonstration Cities program. The second condition was growing recognition of the need for remedial work to deal with shortcomings in existing federal programs designed to aid the cities.

Frieden and Kaplan (1975) also emphasize the effect of urban rioting on Demonstration Cities legislation. They observe that the riots in Watts (1965) and in Chicago and Cleveland (1966) "strengthened, on the whole, the sense of legitimacy and urgency" (pp. 54-55) attached to the use of federal assistance in aiding poor city dwellers. On the other hand, if Washington turned its back on the needs of the poor and urban minorities, it would threaten "to tear apart the fragile social fabric that was holding the cities together" (p. 34).

This argument has been termed the *danger-of-violence rationale* for supporting urban legislation such as Demonstration Cities. It came to play a powerful role in legitimizing and shaping the Johnson administration's strategy for dealing with the riots in American cities (Haar, 1975). An escalating siege mentality was taking root in Washington as urban unrest continued to spread. Bornet (1983) describes this syndrome. The enactment of civil rights legislation throughout the 1960s, he points out, corresponded to the assassinations of Robert F. Kennedy and Martin Luther King, Jr., as well as to the violence in several cities. Bornet concludes that adoption of civil rights legislation was stimulated by civil disturbances.

The House Subcommittee Watershed

If Senator Muskie was to put his name behind Demonstration Cities, it was important to find legislative compromises that would sit well with constituents back home. But Maine voters were not the only forces to be reckoned with. Muskie would have to find accord between those who favored immediate action and a high-visibility program versus those who wanted a more thoughtful, effective, and long-range campaign to get at the root of urban problems (Haar, 1975). Increasingly, there were fears that Demonstration Cities catered more to riot responsiveness than to fundamental urban reforms. If this was so, however, others could take heart in the possibility that the president's new initiative might buy time for Democrats to fashion a more penetrating effort in the next session of Congress.

Muskie's staff sought a compromise that would ease the bill's financial threats. It recommended that Muskie tell the president to limit participation in Demonstration Cities to as few as 25 major cities. Only planning grants would be awarded initially. Cities whose plans were successfully reviewed by HUD would then become eligible to receive a package of assistance from existing federal categorical grant programs. The White House and HUD would set up a cabinet-level task force to coordinate the various bureaucracies and secure their cooperation. The president would appoint a federal coordinator and a coordinator in each city. The Muskie staff proposal would eliminate the supplemental grant mechanism and require only a small federal commitment of new funds for planning assistance (D. Nicoll, personal communication, June 15, 1966).

As the crucial Housing Subcommittee vote approached, Demonstration Cities proponents maneuvered to scrape up enough votes to move the measure, H.R. 15890, to the full Committee on Banking and Currency (Semple, 1966e). With direction from Larry O'Brien, White House and HUD staffers concentrated their energies on subcommittee Democrats who were wavering. At that point, five Republicans opposed the bill and three

Democrats supported it. The remaining four Democrats on the subcommittee were as yet uncommitted, but their votes could determine the success or failure of the president's campaign.[6] One potential stumbling block was Subcommittee Chairman Barrett. Known behind his back, for, among other reasons, his silly toupee, as "Silly Williem," Barrett was an old-line Democrat who survived elections by delivering pork to his heavily blue-collar Philadelphia area constituents. Federal programs such as Urban Renewal had been used to revitalize the downtown and city officials wanted nothing to disrupt the flow of money from Washington. Realizing this, Secretary Weaver told participants at the annual convention of the USCM in Dallas that he would push for set-aside funds for additional Urban Renewal activity. O'Brien obtained White House support for an amendment to the bill authorizing an extra $600 million for Urban Renewal projects ear-marked exclusively for use in Demonstration Cities neighborhoods. This feature reassured Barrett that existing Urban Renewal program funds would not be siphoned off to Demonstration Cities areas (Semple, 1966k).

By the subcommittee vote on June 22, proponents felt that they had the necessary votes. After weeks of behind-the-scenes maneuvering, the Demonstration Cities bill was approved 7 to 3 by the unit. Abstaining was Representative Paul A. Fino, a Bronx Republican, whose views would surface in the weeks ahead. Climaxing a 5-month effort, the subcommittee's decision gave new hope to urban supporters. But some concessions were made along the way. Chief among them was the original $2.3 billion price tag proposed for financing approved local Demonstration Cities programs. Although the subcommittee retained the original $12 million for planning grants, it sanctioned open-ended funding, leaving to the full Committee on Banking and Currency or the full House the choice of a total budgetary figure. Supporters reasoned that this would allay fears that the program would become too costly. It would also allow flexibility in the House and Senate to raise or lower the funding depending on the state of the nation's economy and federal revenues. Subcommittee members running for reelection were thus less vulnerable to being pilloried by their opponents as big spenders.

Finally, the supplemental grant formula was slightly altered. Instead of giving every successful Demonstration Cities recipient an amount equal to 80% of the local share of total project costs, an amount "up to 80%" was substituted. This had the effect of giving HUD greater discretion to determine the actual supplemental funding each participating community received. The legislation stipulated that HUD would examine population density, juvenile delinquency, unemployment, and other indicators of need in finalizing the amount (Semple, 1966e).

Muskie Accepts at Last

Spurred on by the House Subcommittee vote, President Johnson ordered Califano to meet with Muskie and nail down his commitment to sponsor the bill in the Senate (Haar, 1975). A meeting followed but little progress was made. Revisions recommended earlier by Muskie—such as limiting funding to planning grants—were now of little consequence given the version adopted by the House subcommittee. But the senator took courage from the fact that the measure had surmounted the subcommittee stumbling block. Whatever could be said about his personal interest in urban problems, Muskie could not overlook the fact that Demonstration Cities was now a more viable instrument than it had been only a few weeks before.

The July 4th congressional recess was approaching and Muskie had plans to spend it with his family at his summer home in Kennebunkport, Maine. In characteristic fashion, Johnson ordered Califano to meet with Muskie again. Before he left for the airport the next day, Muskie joined O'Brien, Califano, and Vice President Humphrey at a White House lunch and talked over their differences on the legislation. It was agreed that over the recess, Califano aide Larry Levinson, Phillip Hannah of the Budget Bureau, and Don Nicoll would modify language in the bill in an effort to find common ground. Among the changes they worked out, smaller cities—of the type found in Maine—were made eligible for participation in Demonstration Cities. It was clear to Muskie and others that even though Demonstration Cities had been conceived as a big-city bill, it would never survive the legislative process without opening eligibility.

Meanwhile, on June 28, the full Banking and Currency Committee approved the subcommittee bill 18 to 8 with 7 abstentions. The *New York Times* noted that only a few "insignificant changes" were made in the text of the bill (Semple, 1966d). Yet even though the measure had survived beyond the hopes of most observers, it still faced a Promethean challenge in the weeks ahead.

Muskie was aware that Johnson viewed the Demonstration Cities bill, perhaps first and foremost, as a measure to reduce the likelihood of further urban riots. In later communication with Califano, Muskie's staff reiterated this belief (D. Nicoll, personal communication, June 29, 1966). The proposal, the memo said, was meant to show how the government could coordinate aid for urban physical and social revitalization. It was also meant to associate the president with an effort to deploy existing federal resources to help citizens improve the quality of life in their communities. But the memo prefaced these goals with what some observers suspected was the highest priority for the administration. The president's proposal was meant to "prevent, if possible, a repetition of the Watts-type riots in the 20-25 crisis cities, this summer." Califano, O'Brien, Hannah, and Muskie aides labored over revisions to the House version of the bill through the Fourth of July weekend. But Johnson was growing ever more restless and ordered Califano and O'Brien to Kennebunkport to reach closure with Muskie over the bill (Haar, 1975; Semple, 1966k). Within a few days, the Maine senator had agreed to sponsor the bill in the Senate.[7]

With the crucial House subcommittee and committee votes successfully completed, the measure gained new momentum. In an effort to capitalize on the momentum, John Gunther, executive director of the USCM, sought new endorsements from other Washington urban interest groups. Until then, many had sat on the fence, some expressing positive views about the desirability of Demonstration Cities but having reservations about the proposal as written. Gunther organized a series of meetings beginning July 9. Known informally as the Urban Alliance (Haar, 1975), the group met at the Continental Room of the Statler Hilton Hotel.[8] For the first time, some of the critical interest groups representing cities, mayors, counties, unions, and HUD staff were sharing their concerns and working toward a consensus (D. Nicoll, personal communication, July 11, 1966).

In response to issues raised at these meetings, Muskie's staff recommended to Califano that HUD work with the various interest groups to prepare some illustrative examples of the types of city problems to which Demonstration Cities could respond. He also suggested that other federal agencies be involved in the meetings to develop their support for the proposal (D. Nicoll, personal communication, July 11, 1966). The mood among Demonstration Cities advocates was lukewarm optimism tempered by memories of Harlem, Rochester, Philadelphia, and Watts. As the summer progressed, however, aspirations for a violence-free year in the nation's cities would disappear.

Notes

1. Not since a similarly dramatic performance in the late 1930s by opera singer Marian Anderson, denied access to the Daughters of the American Revolution Hall in Washington, had the federal Mall experienced such an outpouring of sympathy for the plight of blacks in America.

2. Others on the new task force included Edgar G. Kaiser, president of Kaiser Industries; William L. Rafsky, executive vice president of the Old Philadelphia Development Corporation; Walter Reuther, president of United Auto Workers; Senator Abraham A. Ribicoff (D-CT), a former secretary of the Department of Health, Education, and Welfare; Whitney Young, executive director of the National Urban League; Kermit Gordon, president of the Brookings Institution and former director of the U.S. Bureau of the Budget; Charles M. Haar, Harvard Law professor, who would soon join HUD staff as assistant secretary for metropolitan development; and Ben W. Heineman, chairman of the board of the Chicago and Northwestern Railroad.

3. Apparently, a turning point came when Larry O'Brien, Johnson's postmaster general, rallied key figures in the Model Cities campaign. O'Brien, HUD Secretary Robert Weaver, and others met on June 10 in Califano's White House office. A gloomy mood about the Demonstration Cities proposal pervaded the atmosphere; only Weaver wanted to push vigorously for its enactment. Sensing the dispirited attitude, O'Brien rose and chided the group, saying that he had never heard of Johnson appointees conceding defeat on any measure the president really wanted. O'Brien then challenged the group to secure passage for Demonstration Cities. Knowing the high regard in which the president held O'Brien, the group pledged to forge ahead (Semple, 1966k).

4. By this time, Muskie, who was first elected to the Senate in 1958, had accrued a promising record of sponsorship for modest legislative proposals but had never managed a major piece of legislation. The Johnson White House knew that Muskie, coming from a predominantly rural state, had less to gain personally or politically from such a bill. This relatively disinterested perspective, coupled with growing respect in the Senate for Muskie's integrity, might improve Demonstration Cities' prospects for approval.

5. All references to personal communication pertain to materials in the Files of the Edmund Muskie Archives at Bates College in Lewistom, Maine (Box 593, Folders 1-5).

6. Representative Leonor K. Sullivan (D-MO) became the target of Andrew Biermiller, head lobbyist of the AFL-CIO, who was recruited by HUD Secretary Weaver to try to persuade Sullivan to vote yes. With many union members in her home district, she eventually saw the light and pledged her support (Semple, 1966k, p. 5).

7. Finishing their work on July 7, Johnson's aides were flown to Pease Air Force Base in New Hampshire, braving foul weather and harrowing flying conditions (Califano, 1991). From there they motored up to Muskie's Kennebunkport home. Heartened by the House subcommittee vote and revisions to the bill, Muskie warmed to the task asked of him. Over Jane Muskie's celebrated lobster stew, the entourage succeeded in gaining the senator's pledge of sponsorship.

8. The first gathering brought together HUD Secretary Weaver and Undersecretary Wood; Andrew Biermiller of the AFL-CIO; C. D. Ward of the National Association of Counties; Patrick Healy, head of the NLC; Jack Conway of OEO; and others. As many as 60 people attended, 15 from HUD staff.

4

Reverend King, the Urban Poor, and an Epidemic of Rioting, 1966

The summer of 1966 proved to be a broiling one. Memories of the Harlem and Watts disturbances were still fresh in the minds of many Americans. Meanwhile, civil rights organizations such as the Southern Christian Leadership Conference (SCLC) and the Student Nonviolent Coordinating Committee (SNNC) continued their sit-ins and demonstrations in southern communities. But the SCLC had decided that it should also extend its campaign for black civil rights from southern communities such as Atlanta, Birmingham, and Selma to the crowded slums of Northern cities. It was a critical strategic move on the part of the SCLC because its primary base of support was among southern black church congregations. But the rising tide of violence in Harlem, Watts, and Philadelphia and the challenge of northern church officials and civil rights leaders convinced King and his followers that it was time to nationalize their effort. For the first time, the black victims of poverty and discrimination in northern cities such as Chicago had leadership and a voice.

Doubtless, few, if any, of King's colleagues in the SCLC could remember the Chicago riots of almost a half-century earlier. But the legacy of racial divisiveness demonstrated by those long-ago conflicts was well known to many. In addition, Chicago had acquired something of a reputation among civil rights workers. The Windy City, with a vast population of poor blacks, was highly segregated. Blacks generally lived on the bottom rung of the ladder of progress. With the local government firmly in the control of Mayor Richard Daley's political machine (the last big-city political machine in the country), Chicago became the stage on which the next scene in the civil rights movement would be played out.

Chicago

As King and his followers mobilized, a relatively minor altercation between police and black citizens not far away ruptured into open lawlessness. With temperatures near 100 degrees on Tuesday July 12, black youths in the Near West Side opened two fire hydrants to cool off. Police arrived and turned off the flow of water. Soon, gangs of up to 250 children and teenagers were roaming the area. Rocks, bottles, and Molotov cocktails were thrown at police cars. As the hours passed and heat waves rippled from the asphalt, crowds composed largely of black youths began looting stores. At least five shots were fired by police. King, who had appeared at a civil rights rally in the city 2 days earlier, talked police into releasing six arrested black youths into the custody of a local minister ("Negro Youths Attack Police," 1966). The next day, the pattern of lawlessness was more or less repeated and two blacks were admitted to a hospital with gunshot wounds.

King told Mayor Daley that an immediate problem was the lack of enough swimming pools and recreational opportunities for the city's poor areas. A deeper issue was police-citizen relations, he said. King called for a civilian police review board such as that of New York. Daley, however, was noncommittal. By dusk, 400 police had mustered in the Near West Side. Three police vehicles were set afire, tracer bullets were fired from a public housing project, and three buildings were torched. King continued to call for calm. As if the rioting were not enough, eight nurses had been found murdered in a hospital dormitory earlier that day in what became one of the most infamous crimes of the decade. Their killer, Richard Speck, would soon be arrested and the tragedy would blemish Chicago's reputation for years afterward (Wehrwein, 1966).

Although temperatures had dropped by 25 degrees, more violence followed on July 14. Gangs of blacks moved along Roosevelt Road and side streets, smashing windows, heaving firebombs, and hooting and retreating when police arrived. The following day, Governor Otto Kerner called up 4,000 National Guard troops to patrol the riot area with carbines and fixed bayonets. By nightfall, violence had virtually ceased. But not before two blacks—a 14-year-old girl and a 28-year-old man—were killed and six police officers were wounded by snipers. In addition, 51 others were injured by gunshots, broken glass, rocks, and other objects. Police arrested 282 people that day (Janson, 1966). In less than 2 years, Governor Kerner's name would become synonymous with a presidential commission he chaired that was set up to study the causes of rioting and recommend appropriate federal responses.

In a conciliatory gesture, Daley agreed to install sprinklers on hydrants so that youngsters could cool off on hot days. Undeterred, King attributed the violence to segregated living conditions and police brutality. Daley retaliated, claiming that SCLC aides to King had incited youths to riot by showing films of the Watts riots. King parried, arguing that the movies illustrated the negative side of rioting (Janson, 1966).

Cleveland

Officials in Washington and in cities all over the nation watched anxiously as the long hot summer wore on. Within days, another Great Lakes city was on fire. A march on Cleveland's city hall to protest discrimination in job hiring had proceeded peacefully on a muggy July 18. About 200 blacks participated. Nevertheless, at around 8 p.m. rioting broke out at Hough Avenue and 75th Street, a predominantly black neighborhood. A 26-year-old mother was killed by gunfire, two black men were wounded, and four policemen were injured by thrown objects. Within 3 hours, gangs roamed the Hough Avenue corridor and side streets setting fires, looting shops, and breaking windows. Firefighters came under gunfire ("Negroes Riot in Cleveland," 1966).

Meanwhile, New York Mayor, John Lindsay, recalling similar violence in Harlem and the Bedford-Stuyvesant section of Brooklyn two summers before, walked East Harlem's 118th Street with his aides. The popular mayor, tie loosened and shirtsleeves rolled up, had been making unannounced appearances in minority neighborhoods to show sympathy and concern for residents trapped in the blistering heat. "Anything that will keep the peace in the ghettos," Lindsay declared softly, "is well worth the effort" (Smith, 1966, p. 1).

Mayors were not the only ones who grew increasingly worried by the violence in the Midwest. The White House chose to adopt a "good cop, bad cop" stance. Vice President Hubert Humphrey, appearing at a meeting of the National Association of Counties in New Orleans on July 18, reminded his audience of the underlying frustrations of ghetto residents. Humphrey allowed as how he, too, might lead a pretty good revolt "if rats nibbled" at his children's toes. He also cited filthy streets, uncollected garbage and a lack of swimming pools and recreational facilities in poor areas. The vice president urged that city slums be wiped out and that new planned communities be built. Many interpreted his comments as sympathetic to the urban violence. It could not escape notice, too, that Humphrey termed these actions *revolts,* connoting a degree of legitimacy, rather than *riots,* suggesting destruction for its own sake ("Humphrey Warns of Slum Revolts," 1966).

In a bad cop stance a few days later, President Johnson, appearing before the Indianapolis Athletic Club, revealed a less accommodating side of the White House. In what the *New York Times* called his strongest words since the Chicago and Cleveland outbreaks began (Pomfret, 1966), the president condemned the perpetrators and warned that they would suffer the heaviest consequences. Moreover, he added, the lawlessness would only turn away those who most wanted to improve conditions for blacks.

On the second day of violence in Cleveland, Ohio Governor James A. Rhodes declared a state of emergency and mobilized 1,500 National Guard troops. Over the next 4 days, the unrest continued, but large mobs formed less frequently and property damage diminished. Still, sporadic sniper fire occurred, stones and bottles were thrown at those patrolling the area, and stores were pillaged (Rugaber, 1966b).

By July 21, Cleveland's death toll had reached four, including a 29-year-old black man shot in his car in a white neighborhood adjacent to the predominantly black East Side (Rugaber, 1966a). Although temperatures had dropped, unrest persisted. The city's chapter of the National Association for the Advancement of Colored People (NAACP) called for racial integration of the police force and appointment of a citizen's committee to investigate the riots and make recommendations. Both city officials and the local NAACP concurred that although some of the violence was spontaneous, much was planned and organized. The head of one local black militant organization declared that blacks would "have to take whatever action is necessary to get them their rights" (Rugaber, 1966c, p. 1). This leader drew on a metaphor for the Vietcong used among the troops in Vietnam. He warned that, as Cleveland blacks gained experience, the disturbances could spread to "Charlie's [the white man's] front door" (Rugaber, 1966c, p. 1).

The Muskie Substitute Bill Emerges

Back in Washington, the explosive events in the Midwest sparked a renewed urgency for passage of the Demonstration Cities bill. Yet it was also clear to everyone that, whatever its potential for ameliorating future urban mob actions, hope was gone that the bill could be enacted in time to have any effect in 1966. With congressional elections in November, members of the House in particular tended to exercise a short-term view as they examined legislation. In spite of this, two forces conspired to press the bill forward. First, the outbursts in Chicago and Cleveland confirmed for many that earlier episodes of unrest in Harlem, Watts, and other cities were harbingers of what now appeared to be a pattern in urban America. (As if to reinforce this perception, Philadelphia became the scene of yet another riot later that summer.) As black militancy increased, the promise of continued violence became manifest. One study points out, that these conditions "strengthened, on the whole, the sense of legitimacy and urgency" (Frieden & Kaplan, 1975, pp. 54-55) attached to the use of federal aid to help the urban poor.

A second force giving Demonstration Cities momentum was the fact that the measure appeared to strike directly and comprehensively at the conditions thought to cause rioting. Because of this riot responsiveness and because the bill had already advanced partially through the circuitous web of legislative channels on Capitol Hill, Demonstration Cities was slowly emerging as the most promising option available to members of Congress searching for a vehicle to deal with the riots.

While the destruction unfolded in Chicago and Cleveland, Muskie's staff rewrote the Demonstration Cities bill. The so-called Muskie substitute version was completed in the waning days of July. Muskie refined the measure and made it more palatable to members of the Senate without changing the essence of earlier versions. The Muskie substitute clarified language, sharpened the objectives of the proposal, corrected certain oversights that might have distorted the program's operation, and placed greater emphasis on local creativity in designing programs. Furthermore, it toned down sections providing for housing desegregation, making them less explicit and leaving more discretion to the Department of Housing and Urban Development (HUD) (Haar, 1975). It also mandated cooperation among participating federal agencies, with HUD in the coordinator role. In addition, it made clear that the program was an experiment and would terminate by June 30, 1971.

Muskie's most substantive recommended changes were in the supplemental grant provision. Under his version, localities could continue to apply for these grants to help pay the portion of costs for local projects not covered by other federal grants. Up to 80% of the local share could be paid out of supplemental funds.

Muskie recognized that most cities would try to "attach" as many federal categorical grants to the Demonstration Cities neighborhood as possible, thereby inflating the base on which their supplemental grant would be calculated. Consequently, he added language to his version of the measure that would exclude irrelevant projects from supplemental grant computations. Moreover, when cities had agreed to pay the local matching share of project expenses on grants secured before their designation as a Demonstration City, Muskie's version would prevent them from shifting this liability to the supplemental grant. It also tightened language prohibiting city halls from using supplemental grants to finance general city administration costs. Another addition required cities to place higher priority on using supplemental grant monies for innovative nonfederally aided projects before using them to help finance conventional federally supported projects. Finally, to reassure mayors in cities with existing Urban Renewal projects outside the Demonstration Cities neighborhood, the senator added wording to prohibit the government from diverting existing levels of funds from that program for use in the Demonstration Cities area.

Endorsed by the Senate Subcommittee

The Senate Housing Subcommittee, having held hearings in the spring on the Johnson administration version of Demonstration Cities, took up the Muskie Substitute on July 23. Muskie cast the new measure in its best light. On his side were the earlier House subcommittee and committee votes. Some Senate subcommittee members were impressed that Muskie apparently had little to gain from Demonstration Cities. For even though eligibility to participate in the program was no longer limited to the largest cities, it appeared that Maine—with few minorities and no riots—was an unlikely contender for designation. Moreover, numbers, as usual, played an important role. In this case, it was not as critical to the 100 senators (as it had been to the 435 representatives) that only 50 or 60 cities would benefit from Demonstration Cities. Each of Muskie's colleagues had hopes that at least one community in his or her state would be selected to participate. Representatives, on the other hand, knew that most of their districts would not be winners in the Demonstration Cities sweepstakes.

Another numerical issue involved dollars. Some members of the Senate subcommittee were worried about spending and looked askance at the proposed $2.3 billion budget to be paid out over 5 years. One senator in particular, Thomas McIntyre—Muskie's fellow Democrat from neighboring New Hampshire—worried that if he supported the bill, he would be labeled a big spender among tightfisted voters during his upcoming reelection campaign. Learning this, and knowing that McIntyre was a key vote on the Senate subcommittee, the president agreed to let the New Hampshire Democrat propose an amendment reducing the budget from $2.3 billion over 5 years to $900 million over 2 years. Muskie and his staff prepared the amendment and quietly passed it on to Muskie's New Hampshire colleague during the subcommittee meeting. As McIntyre announced his addition to the bill, it was clear that senatorial comfort levels were inching upward. On July 26, in what might otherwise have been a tie vote, the subcommittee approved the bill 6 to 4, thanks to McIntyre's support. For his efforts, McIntyre was then in a position to tell voters that he had successfully spearheaded an effort to reduce government spending by $1.4 billion (Califano, 1991; Haar, 1975; "Slum Aid Slashed by Senate Panel," 1966).

Although this was another important step forward, the event left the bill's financial muscle much atrophied, raising questions whether much could be demonstrated by Demonstration Cities. Two alternatives lay open to supporters of the bill: get the full House and Senate to agree on a larger appropriation or reduce the number of cities to which awards would be made. The former was guaranteed to offend fiscal conservatives; the latter was certain to discourage potential supporters whose vote depended on reassurances that their state would get at least one Demonstration Cities award.

Letters and telegrams of support for the bill trickled into Muskie's Senate office. He received favorable correspondence from officials of the National Conference of Catholic Charities (personal correspondence, August 1, 1966) and the American Public Welfare Association (personal correspondence[1], August 10, 1966). The head of the National Association of Counties pledged support but asked that the measure be renamed the County-City Demonstration Program and that counties be eligible to participate (personal correspondence, August 12, 1966). As support continued to fall into place, the full Senate Housing, Banking, and Currency Committee met in early August, confining most of its revisions to restrictions in the supplemental grant mechanism. On August 19, the committee adopted the measure 8 to 6.

While the congressional engine bearing Demonstration Cities groaned forward, the Johnson administration continued to link the initiative to the danger-of-violence argument. On August 17, Attorney General Nicholas Katzenbach, speaking in New York City, warned his audience that urban tensions could explode that summer in as many as 30 or 40 cities, bringing violence and destruction. He chided Senator Robert Kennedy (D-NY), who lambasted the Johnson administration for having no program in place to address the ills of the cities. Katzenbach, a friend of the Kennedy family, reminded the senator that several programs were in place and others, such as Demonstration Cities, were waiting in the wings. Kennedy scoffed, calling these programs "just a drop in the bucket for what we really need" (Hunter, 1966a, p. 31). Katzenbach, in the words of a *New York Times* reporter, said that Demonstration Cities was "essential if the nation was to wipe out urban blight and erase the causes of rioting" (p. 31). It was becoming apparent to Johnson aides that Kennedy was setting himself further apart from the president, hoping to capture the support of an urban constituency in preparation for a run at the White House.

Deliberations on the Senate Floor

On August 19, the full Senate met to consider S. 3708, the Demonstration Cities legislation. In his introduction of the measure, Muskie invoked images of youth and innocence. He reminded his colleagues that the bill could help "achieve the American dream for the child whose playground is a trash-strewn alley, whose classroom is a rat-infested cellar" (Hunter, 1966c, p. 2). Abraham Ribicoff (D-CT), a member of the Wood task force that had conceived of Demonstration Cities, immediately cast a cataclysmic mood in the Senate chamber, asserting that the urban crisis "concerns the very future of our democracy, and, in great measure, will determine the future course of the democratic ideal throughout the world." The nation was at a crossroads. Ribicoff asked rhetorically whether cities could continue to function "in a democratic sense" or whether they would "sink in a welter of economic disuse and social disorganization." He lamented the failure of previous federal efforts to grapple successfully with urban problems and pointed to "headlines of riots, of civil disobedience on a vast scale" in the "festering slums of the cities" (*Congressional Record,* 1966a, pp. 20061-20063).

Another supporter, Senator Paul H. Douglas (D-IL), gave a thoughtful account of the causes and consequences of several urban problems. Mindful of the Chicago riots only a month earlier, Douglas stressed that "time is of the essence" in dealing with these matters. He expressed confidence that the bill "will demonstrate to the nation that we can rescue our cities" (*Congressional Record,* 1966a, p. 20064). Douglas was followed by others whose support was cast in less antediluvian terms (pp. 20065-20067). But the danger-of-violence flame was rekindled by Gaylord A. Nelson (D-WI). He cited a litany of urban problems and declared that the "future is not bright." Frustration, "born of despair," had "led to riots and violence: We must act and act now, for the crisis of the cities is a nationwide crisis" (p. 20068).

Joining Douglas was another riot-weary senator from the Midwest, Stephen M. Young (D-OH). The devastation in Cleveland's Hough area was caused, Young announced, by the "indignities of ghetto life—the pent up frustrations, unemployment and hopelessness of many persons living in crowded and neglected city slums." He lamented that if monies devoted to foreign aid had instead been directed to doing away with "slums and ghettos and to aid high school dropouts and provide employment, recent rioting in city slums would not have occurred because the worst slums would have been eliminated." The urban crisis, he insisted, was "one of the most profound internal problems with which our country has had to wrestle since the days of the great depression" (*Congressional Record,* 1966a, p. 20060). Young deplored the violence in American cities but was convinced that more would come unless Congress took action (*Congressional Record,* 1966a, p. 20070). He closed by urging his colleagues to stamp out the slums that had given rise to the rioting wracking the nation's cities.

Although many of those who opposed the measure had already communicated their views, conservative Everett M. Dirksen (R-IL) felt compelled to summarize their positions. He thundered that the program "has the prospect of becoming one of the greatest boondoggles this country has ever known." He fumed that Demonstration Cities would be "shot through with waste and corruption and goodness knows what all before we get through" (Hunter, 1966c).

Underlying objections to Demonstration Cities among many conservatives—and some moderates—was a clause that would require new federally funded housing in Demonstration Cities neighborhoods to be racially integrated. HUD Secretary Weaver, himself black, had warned the president earlier that, "there are overt and hidden implications of racial integration in the proposal" (Lemann, 1991, p. 197). As other concerns were lodged over the summer, the integration section was deleted, doubtless easing resistance to the bill.

On August 19, the Senate vote was finally taken. The measure carried by a comfortable margin, 53 to 22. The budget had been boosted from the subcommittee's $900 million recommendation to $1.2 billion, slightly more than half the amount Johnson had originally requested. The program would run over 3 years rather than the 5-year period Johnson wanted. And to mollify Urban Renewal supporters, the bill included an additional $250 million earmarked for that program to be used on projects in Demonstration Cities neighborhoods.

When the Senate-approved measure was sent back to the House for reconsideration in late August, it was combined with another bill providing support for metropoliswide planning and was labeled the Demonstration Cities and Metropolitan Development Act.

Back in the House

Although the original version of the Demonstration Cities bill had been introduced in the House by the president's colleague and friend Wright Patman, the congressman had done so without much apparent enthusiasm. Having survived the first House subcommittee vote and the full Senate and with growing support among various constituencies, however, the much-revised measure took on new appeal to the Texas Democrat. With his cosponsorship, Demonstration Cities (and its companion Metropolitan Development section) had a critical ally. The House Subcommittee on Housing accepted the Senate version with few changes. Approved on August 25, the bill was sent on to the full committee. There, the Demonstration Cities proposal became Title I and Metropolitan Development became Title II of an omnibus bill. The other titles incorporated numerous modifications to various housing and historic preservation programs. Reported favorably out of committee on September 1, the measure emerged with Demonstration Cities essentially untouched (Haar, 1975). Omnibus bills had become a time-honored tradition in Congress, making it possible for individual members to vote for a measure even though they may not agree with every provision. One's vote could then be explained to critics as a necessary evil to deliver the positive features of the legislation to one's constituents (Haar, 1975).

As the summer wore on, proponents of the president's measure continued to fall into line. Department of Health, Education, and Welfare Secretary John Gardner told a Senate subcommittee examining urban problems that his department supported Demonstration Cities. This was a signal to senators that at least some support for the program existed within the federal bureaucracy. In the event of passage, HUD would call on several agencies to chip in categorical grant funds to help support Demonstration Cities neighborhood programs in education, job training, day care, income maintenance, employment referral, and the like. Although many in the federal bureaucracy were not enthusiastic about a program that might draw resources away from their constituencies, Gardner's testimony was an indication that intragovernmental cooperation was a possibility (Hunter, 1966b).

Meanwhile, Johnson got wind of a plan among House Republicans to kill the bill. He directed his aides to press newspapers and businesspeople to endorse the bill (Califano, 1991). A week before the full House considered the bill, a *New York Times* editorial urged Congress to approve the measure, calling it a "start in the right direction" ("To Rescue the Cities," 1966, p. 46). That same day, the president spoke out, calling Demonstration Cities "one of the most important pieces of legislation for the good of all American mankind that we can act upon this session" (Califano, 1991). Three days prior to the House session, a group of 22 Fortune 500 businesspeople had publicly endorsed the measure. Among them were David Rockefeller, Henry Ford, and Robert Lehman. Also included was Edgar F. Kaiser. Kaiser had been an influential member of the task force that crafted the original Demonstration Cities proposal (Semple, 1966c).

On the eve of the House's vote, the *New York Times* returned to the issue with another editorial, this one giving even more forceful support to Demonstration Cities. "No other measure that has come before Congress in this session is more significant," it announced ("Demonstration Cities," 1966, p. 44). The paper scolded Representative Paul Fino (R-NY) for employing "scare tactics" in his opposition to the metropolitan section of the bill. The Bronx Republican had charged that Title II would promote racial integration in public schools. In truth, Title II was intended simply to spur coordinated comprehensive land use planning among cities and suburbs to discourage suburban sprawl.

On October 14, Demonstration Cities was brought before the House for a vote. Johnson's aides had succeeded in mobilizing an impressive cast of characters in furtherance of the bill. With key senators, trade association lobbyists, welfare organizations, state and local elected officials, unions, civil rights groups, corporate heads, and the *New York Times* behind him, the president was in a vastly better position than he had occupied 6 months earlier. Nevertheless, by mid-October the measure could no longer capitalize on the urgency posed by the long, hot summer. Instead, the record of violence that summer was offered as testimony to the importance of federal action in the coming year. In its last major battle, Demonstration Cities generated 8 hours of debate preceding the House vote, an indication of the depth of concern about the bill. Three attempts were made to stymie its passage or reduce its provisions.

Money and Racial Animosity
on the House Floor

To allay fears that the measure could be used to foster greater racial integration, Representative Abraham J. Multer (D-NY), a Brooklyn Democrat, succeeded in adding language to both Titles I and II assuring that nothing in those sections could be construed to this effect (Semple, 1966b). The particular provision that troubled some in the House was one of the 14 criteria to be used by HUD in selecting the Demonstration Cities grant recipients. This would give credit to proposals that sought to "counteract the segregation of housing by race and income." Ultimately, this passage also was eliminated in the House bill.

At the center of the race issue was Bronx Congressman Fino again. He had been unsuccessful in the House subcommittee and committee deliberations in his attempts to kill the measure. Consequently, he was suspected by some of raising a red herring in a last-ditch attempt to diminish, if not defeat, the bill. He opposed the Multer amendment, calling it a "pipsqueak." S. 3708, he said, was "one of the most dangerous" pieces of legislation "ever drawn" (*Congressional Record,* 1966b, p. 26922). Despite assurances from Secretary Weaver that neighborhood schools, their attendance districts, and bussing would not be affected by the legislation, Fino held his ground. Even the Multer amendment was not enough to move him. Although Fino was in the minority in the House chamber that day, a few representatives, such as William T. Cahill (R-NJ), also expressed fears that forced bussing lay beneath the surface of S. 3708.

Tennessee Representative William E. Brock, III, proposed an amendment to remove $900 million of action money, leaving $24 million to allow cities to prepare plans only (*Congressional Record,* 1966b, p. 26927). This would force HUD to return to Congress later to request implementation funds. House sponsor Wright Patman immediately opposed the amendment, saying it reached "for the jugular vein to destroy" (p. 26927) the bill. Continuing to play on racial fears, Fino spoke on behalf of the change. There was "a very real danger," he declared, "that black power is going to ride the Demonstration Cities gravy train in some cities" (p. 26929). Then the Bronx congressman fired his heavy artillery. He noted that San Francisco's Urban Renewal director, in anticipation of the measure's enactment, was "dickering with black power for control of the local Demonstration Cities Program" (p. 26929). This would bring "demonstration rioting and demonstration anarchy" (p. 26929), he fumed. Fino revealed that he had written the president asking him to hold up the measure until the Justice Department or the House Un-American Activities Committee could investigate whether a "black power takeover" (p. 26929) was in the works. Fino charged that Weaver continued to deny that the Demonstration Cities program would foster black power.

Perhaps sensing that his cause was ill fated, Fino decided to pull out all the stops, delivering even more outrageous accusations. Weaver, he charged, had "bragged" that Demonstration Cities would remake school systems and could be used to deny program funds to cities that did not adopt open housing and civil rights ordinances. The law could also be used, he continued, to encourage communities to revise their zoning ordinances and property tax policies and adopt subsidized rental housing programs in residential neighborhoods, "like it or not" (*Congressional Record,* 1966b, pp. 26929-26930). The bill, Fino railed on, was "just a gimmick to centralize power" and to "promulgate zebra-colored housing and education guidelines." He said that it could become "a gravy train for black power," insisting that it was "a half-baked racial balance scheme that may well be a bankroll for black power" (p. 26930).

If the Bronx congressman was the most ardent opponent of the bill, he was not the only detractor. Some, such as James Harvey (R-MI), were merely disappointed by the potential effect of Demonstration Cities, saying too few cities would be involved in the program. Harvey complained about the $900 million price tag, warning that as many as 700,000 troops could be in Vietnam by 1968. In the future, the war could strain the federal budget even further, he admonished. (Within a year, in fact, U.S. troops would be patrolling the riot-torn streets of Detroit.) He supported the $24 million amendment as a way to limit federal expenditures (*Congressional Record,* 1966b, p. 26931). Others in the chamber turned the Vietnam War argument around. Representative William F. Ryan (D/L-NY) belittled the $900 million figure, saying that it was "less than we spend in one week in Vietnam" (p. 26936).

The racial theme surfaced again, this time in a strange and oblique reference by Joseph D. Waggoner, Jr. (D-LA). He reminded his colleagues that he was from a small town where the local Howard Johnson's "has only one flavor and that is vanilla" (*Congressional Record,* 1966b, p. 26935). The *Congressional Record* does not record whether there was a gasp in the chamber in response or whether a more merciful silence followed. But no one dignified what even the most charitable judgment could only term an ill-conceived metaphor for another point.

Continuing the play to white fears, Albert W. Watson (R-SC) warned that S.3708 funding would be used primarily for "the purpose of racial, social and economic integration and not to remove the blight in our cities" (*Congressional Record,* 1966b, p. 26941). In an apparent reference to the recently enacted Voting Rights Act, he carped about "federal registrars" trying "to direct our elections." The measure before him, he argued, would "put an economic pistol to the heads of our cities to be used primarily for further racial integration in housing and in our neighborhood schools" (p. 26941). When, at last, debate was closed on the amendment to limit funding to $24 million in planning money, it failed 141 to 110.

Sensing that the Title I portion of the bill (Demonstration Cities) had enough supporters to secure passage, Fino proposed to strike Title II, the Metropolitan Development section, from the measure. His amendment failed 93 to 63. A third measure attempted to combine both the defeated amendments, thus appealing to a larger potential opposition vote. It would have shunted the bill back to the Committee on Banking and Currency, cutting the $900 million and deleting Title II. It was voted down 175 to 149.

It was not until 10 in the evening of October 14 that the measure was ready for a final vote. Thirty-five amendments had been presented and 20 accepted (Haar, 1975). The House voted to approve S.3708, as amended, 178 to 141. The most difficult stumbling block for the bill, the House of Representatives, had at last been surmounted ("Cities Bill Passes Main Test," 1966). Most of the opposition, to no one's surprise, came from Republicans and southern Democrats. But the momentum of a few progressive Republicans and northern Democrats was enough to counterbalance resistance. Nonetheless, the fact that more than 100 members of the House failed to vote on the bill illustrated its "can't-do-me-any-good" character. The light turnout was also due to the fact that many representatives, with only a few weeks until the November elections, were back in their districts tending to home fires.

If the victory was sweet for O'Brien, Califano, Muskie, Patman, and other advocates, the final word in the House was still not in. Congressman Prentiss Lafayette Walker (R-MI) later inserted a statement into the *Congressional Record* opposing Demonstration Cities. Great Society programs, he argued, paid people not to work. He deplored school bussing to achieve racial desegregation and said that America should bus workers to jobs. (Whatever Walker's motivation for advocating a workers-to-jobs strategy, the idea would take on currency with liberals such as Senator Bill Bradley, D-NJ, a quarter of a century later.) Walker concluded that the House should not "try to comply with the demands of our 'one worlders' who are so immoral they would make the many races created by God into one race" (*Congressional Record,* 1966b, p. 27013). If Walker's views were excessively negative, perhaps those of Representative John J. Gilligan (D-OH) were excessively Pollyannaish. His comments, also inserted into the *Congressional Record* after the vote, note that S.3708 was supported by several groups in his home district of Cincinnati. The measure "provides the means by which neighborhoods may be revived, crime eliminated, [and] poverty defeated" (pp. 27042-27043).

It was clear that Demonstration Cities had the momentum to secure congressional enactment. But with adjournment on Capitol Hill scheduled for October 22, action would have to be accelerated. Three days after the House vote, the approved House and Senate versions of the bill went to a Conference committee, where a $1.3 billion measure was fashioned. The *New York Times* called the hybrid "one of the most important pieces of domestic legislation of the year" (Semple, 1966a, p. 18). It was regarded by the president, the *Times* said, "as the foundation of his attack on poverty and blight in the cities" (p. 18).

After the House-Senate conference committee reported its revised version of the bill on October 17, the bill went back to each house for final endorsement. In a statement immediately after the conference vote, Muskie reassured the president that the essential character of the administration version of Demonstration Cities remained intact. He noted, however, that Senate conferees had wanted to restrict supplemental funds to "new, innovative non-federally aided programs" (personal communication, October 18, 1966). House conferees, on the other hand, proposed allowing it to be used to supplement financing for projects covered in part by other federal programs. Muskie knew that if the House was successful, this measure would erode the effect of the Demonstration Cities legislation, potentially creating a slush fund for conventional pork barrel projects in public housing, Urban Renewal, highways, sewer and water facilities, and the like. Fortunately, the House participants in the conference agreed to rescind their request, thus restoring the bill's original intent.

In Sight of the Finish Line

As the Senate and House each prepared to vote on the conference committee version of the bill, the *New York Times* set the mood for final consideration of the bill, calling it no less than "the key to [Johnson's] attack on urban blight and poverty" (Semple, 1966j, p. 1). On October 18, the Senate voted, awarding S. 3708 a 38 to 22 approval. Two days later, the measure squeezed through the House 142 to 126, with a whopping 160 members abstaining. Supporters included 11 Republicans and 131 Democrats. The Democrats were from all sections of the United States, whereas the Republicans were from five northeastern states: New York, Pennsylvania, Maine, New Jersey, and Massachusetts—most heavily urbanized states. Opponents of the bill included 79 Republicans and 47 Democrats. Similarly, the Republicans were from all over the nation and Democrats were regionally centered—in this case, in the southern states. Thus, a coalition of Democrats and northeastern Republicans was able to defeat a slightly smaller group of Republicans and Dixiecrats by a narrow 16-vote margin.

If racial issues had been prominent in earlier House considerations of the bill, the final encounter raised more concern about costs. Representative Earle Cabell (D-TX) was one of those who expressed reservations. He noted that most of the supporters were from eastern and Midwestern metropolitan areas, "where by their own profligacy they have painted themselves into this corner where their own taxpayers can no longer pay the cost of these past mistakes." These people, he continued, "are asking that the taxpayers of the Nation bail them out of their present predicament" (*Congressional Record,* 1966c, p. 28139) A kinder, gentler view was expressed by Representative Cornelius E. Gallagher (D-NJ), who acknowledged that the legislation was "no panacea" and would not "build a network of utopias" (p. 28139). He felt, thought, that the measure was a beginning and "would gradually make life more livable in the cities" (p. 28139). If not a cure-all, he added, the program could help "avert disaster" and "heal and revitalize our cities" (p. 28139). He closed with the hope that the measure would lay a foundation so that "our heirs in the next 30 years will be able to cope with the problems peculiar to their time" (p. 28139). In the aftermath of the 1992 Los Angeles riot, his sober words bear a fresh poignancy. As I argue in subsequent chapters, if there were any foundations built out of the Demonstration Cities experience, they were primarily foundations of knowledge about federal urban program design and the limits of American voters, presidents, and Congress to sustain long-term efforts to revitalize cities (Frieden & Kaplan, 1975; Haar, 1975; Wood, 1990). It remains to be seen whether these lessons were adequately recalled in the early 1990s.

At a White House ceremony on November 3, President Johnson signed the Demonstration Cities and Metropolitan Development Act of 1966. The *New York Times* recalled that initially the bill was received as a "dud" at HUD and on Capitol Hill. Later, it was "pronounced stone dead, beyond even the resuscitative powers of the president" (Semple, 1966k, p. 1) on at least two occasions during its tenuous progress. It provided, the *Times* said, "a classic illustration of how a piece of legislation can be pushed through Congress" (p. 1). It was no small measure of the new law's controversial nature that Johnson introduced it that day not as Demonstration Cities but rather, as Model Cities. Califano (1991) explains that after the ceremony the president smilingly chastised him: "Do not ever give such a stupid goddamn name to a bill again" (p. 135).

For too many people, Johnson feared, the original name would connote the civil unrest associated with public demonstrations against the
Vietnam War, racism, segregation, and other issues. The name change was
one more sign—if one was necessary—of the unorthodox nature of President Johnson's newest legislative stepchild. HUD Assistant Secretary
Charles Haar (1975) later said of the new program "that Model Cities
passed at all was a surprise. That it was significantly changed in the
legislative process, and in ways that would profoundly affect its future,
was not unpredictable" (p. 91).

Note

1. References to personal communication pertain to materials in the files of the Edmund
Muskie Archives at Bates College in Lewiston, Maine (Box 593, Folders 1-5).

5

Riot-Driven Public Policy, 1966 Through 1968

By the time Lyndon Johnson signed the Model Cities bill, interracial mob violence had occurred in 44 cities, more than 2,200 people had been arrested, 467 people had been wounded, and 9 people had been killed—all in 1996 (Wikstrom, 1974, p. 22, Table 3). In addition to major eruptions in Philadelphia, Chicago, and Cleveland, smaller mobs stormed Detroit, Minneapolis, New York, and San Francisco. Even southern cities such as Birmingham, Tampa, and Atlanta shuddered in reaction to tensions between blacks and police. As members of Congress prepared for November elections, the shifting mood of voters toward cities and racial issues grew more troubling. Contributing to their views was the new role played by the news media.

Urban Violence and
New Ways of Knowing

When episodes of urban rioting occurred during World War II, mass media coverage was limited to print journalism and radio broadcasts. The only moving images of violence available were in brief film footage, such as the Movietone News, at moving picture theaters. Thus, public reaction and political response were delayed and somewhat remote from the events themselves. Television, however, catalyzed mass opinion formation and diminished the ability of politicians to stand apart from the associated issues.

During the mid- and late 1960s, images of storefronts aflame, cars overturned, and looters struggling through broken windows with appliances, food, and liquor became all too common. Police and national guardsmen brandishing weapons, black youths throwing rocks and bottles, crowds chanting curses and threats at police and merchants—these images were becoming unremarkable. The 1960s introduced the American people to urban mob violence in the comfort of their own homes. Nightly coverage of attacks in Vietnam, redneck assaults on southern civil rights demonstrators, and campus building sit-ins at institutions such as Columbia and Harvard contributed to the belief that something was drastically wrong in the United States.

Further propelling the national mood was the growing use of surveys to measure citizen attitudes about public issues such as the Vietnam War, urban violence, and race relations. Sampling techniques grew more sophisticated and mainframe computers advanced processing and analysis of the data. As a result, public officials and the mass media were increasingly able to gain frequent and accurate assessments of public opinion. Polling results also clarified for citizens where others stood on issues that they considered important. Events and opinion formation were linked in a matter of days.

A summation of opinion polls on race and urban issues published a short time after the enactment of Model Cities illustrates the shifting public attitudes on race. This summation (Silver, 1968) found that the share of whites who believed that demonstrations by African Americans helped the cause of civil rights had declined from 36% to 15% between June 1963 and October 1966. Conversely, the proportion believing that demonstrations had hurt the cause rose from 45% to 85%. The percentage of whites who said that their anxiety about personal safety had risen in recent years increased from 45% to 59%. Most whites (80%) and blacks (75%), however, believed that African Americans were peaceful and did not support rioting. Less than 25% of whites felt that blacks desired violence and wished to loot stores. Low-income whites were more likely to doubt that most blacks believe in nonviolence. Unmistakably, public opinion was growing less favorable toward the problems of minorities and the cities. It was in this context that the Model Cities call to arms was sounded.

The Model Cities Legislation

At the heart of the administration's newest urban initiative was a pledge to deal with the most pressing social and physical problems of selected poor neighborhoods, to allow residents and businesses to participate in forging solutions and running programs, and to seek innovative and imaginative ways of reducing poverty and erasing blight and deterioration. This was not only a radical departure in the way Washington normally conducted business with the cities, it was also an enormously ambitious attempt at social change. Included among the allowable uses were efforts

to rebuild or revitalize large slum and blighted areas; to expand housing, job and income opportunities; to reduce dependence on welfare payments; to improve educational facilities and programs; to combat disease and ill health; to reduce the incidence of crime and delinquency; to enhance recreational and cultural opportunities; to establish better access between homes and jobs; and generally to improve living conditions for the people who live in such areas. (Demonstration Cities and Metropolitan Development Act of 1966)

Procedurally, municipalities would select an area of the city with high poverty levels. According to program criteria, a boundary would be put around the area designating it a Model Cities neighborhood. Working with residents and businesses in the neighborhood, the municipality would develop a proposal outlining what measures the city would take to improve the quality of life overall in the neighborhood, estimate the costs, and identify funding sources. If the proposal was among the 75 selected by the

Department of Housing and Urban Development (HUD) in the first year, 1967, the city would receive a grant to prepare a more detailed plan and program to carry out the original proposal. If the city failed to receive a planning grant, it could apply again the following year, when another 75 city proposals would be selected for participation. Neighborhood residents were to have significant involvement in drafting the plan. Once plans were approved by HUD, the city would receive an annual HUD Model Cities action grant, along with grants from other federal agencies and state and local sources. Assuming that the city performed well and made acceptable progress, its funding would be renewed each year for 5 years.

HUD would not only manage the application reviews, it would also review and approve or reject the subsequent plans of the 150 selected model cities. Moreover, it would review annually each city's progress on its Model Cities plan and determine Model Cities funding for the next year. Even more tortuous, however, was HUD's role in trying to funnel the many grants from other federal agencies necessary to carry out each locality's overall Model Cities plan. This coordinating responsibility meant that HUD staff had to meet regularly with personnel from other Washington agencies and try to convince them to fund a given city's Model Cities program.

Because this was a federal demonstration program, however, HUD also had to carry out a fairly elaborate effort to evaluate each city's programs and projects in light of its original goals and the goals of the Model Cities legislation. The assumption behind Model Cities was that Congress would foster widespread experimentation and innovation in a search for new approaches to dealing with fundamental urban problems. From these demonstration efforts, Congress might later choose to design and enact new long-term programs around those that had proven most effective. If, for example, 150 model cities came up with a dozen different approaches to reducing juvenile delinquency, HUD would compare outcomes and costs. From these "laboratory" analyses, Congress would be in a position to fashion long-term programs for use throughout the nation to improve the quality of urban life.

Because most previous federal programs were criticized by social activists as having too little citizen involvement, Model Cities placed considerable emphasis on participation by those living and working in the designated neighborhoods. Yet Congress and the president were mindful of the Office of Economic Opportunity's (OEO) new Community Action Program (CAP), which many elected officials disliked because they were bypassed by CAP workers and volunteers in local decisions involving federal expenditures (Moynihan, 1969). Model Cities sought to balance municipal and citizen power. To prevent city officials from dominating program operations, the new law stipulated that each municipality create a city demonstration agency (CDA). The CDA was to be a separate operating arm of city government with its own budget and its own staff and director, appointed by the legislative and executive branches of the municipality. Although the CDA was a creature of city hall and legally bound to the municipality, it was to be governed by a board that could include residents and representatives of businesses and institutions in the neighborhood, as well as municipal officials and leaders from throughout the community. The board members could be appointed by the city government in consultation with model neighborhood residents or they could be elected by model neighborhood residents under municipally organized procedures (Sundquist & Davis, 1969).

Financially, the programs leaned heavily on HUD Model Cities funding. Up to 80% of planning and administrative costs for approved Model Cities programs could be paid out of Model Cities supplemental funds. The remaining 20% had to be financed from other federal programs and from state and local sources. In addition, programs and projects to be funded with supplemental funds were allowable only if no other federal grant-in-aid monies were available. Thus, supplemental funds were not to be employed to replace funding available from existing federal programs.

The legislation authorized a total of $24 million for cities to develop Model Cities plans in fiscal years (FY) 1967 and 1968 and $400 million to cover operating costs for the 75 cities selected in the first round of applications (FY 1968). In addition, $500 million was authorized for subsequent program expenses in FY 1969. On top of these sources, $250 million was authorized to pay for Urban Renewal projects provided for in an approved Model Cities plan (FY 1968). If Model Cities functioned as intended by Congress, however, these monies would leverage many times their magnitude in categorical grants, loans, loan guarantees, and the like from other agencies. For these, cities would have to apply independently, although HUD would do its best to encourage the agencies to support Model Cities local programs.

Place-Based Policy

One should not overlook a central underlying assumption in the Model Cities program design: Because the perceived problems underlying urban poverty and mob violence were largely geographically circumscribed, Congress targeted those areas alone for remediation. Put another way, because urban racial deprivation was confined mainly to black-majority ghettoes—and because rioting occurred largely in or near the ghettoes— then the appropriate scale of public intervention was the ghetto.[1] This thinking was the cornerstone of a cornucopia of federal urban- and metro-politan-centered assistance programs in existence at the time, all of which confined eligibility to geographically delimited areas. The federal Urban Renewal program, for example, was limited to specific central city project areas that were designated slums or contained "blight." Sewer and water facilities, public playgrounds and parks, economic development, and small business development programs were confined to individual central cities or suburban municipalities or to smaller sections within those jurisdictions. For many such federal assistance programs, geographical targeting was a vehicle for maximizing program expenditures (and, therefore, program benefits) within the area of greatest need. Wealthier city neighborhoods and suburban communities thus could be excluded from participation.

Also associated with most place-based programs was the assumption that federal assistance was a temporary intervention. Once the government subvention was spent, once improvements were made within the targeted district, there would be no need for further subsidies. This notion appears to have served best in support of public capital facilities such as sewer treatment plants, playgrounds, or community health clinics. But place-based programs designed to promote the physical, social, or economic uplifting of an entire urban enclave would later bring into question the "temporariness" assumption. The federal government would reexamine this idea in light of additional experience with place-based programs such as Model Cities.

To Improve the Quality of Life, to Halt the Riots

The response to civil unrest occurring in dozens of cities during 1964, 1965, and 1966 had much to do with the momentum mustered by the Johnson administration and Congress during the Model Cities program's legislative birthing process. Former HUD Assistant Secretary Haar (1975) points out that Muskie and others in Congress and the White House raised the specter of further rioting at numerous points along the way, especially whenever the bill seemed in danger of delay, defeat, or evisceration through amendments. Whether and to what extent the danger-of-violence argument turned the tide and secured passage for Model Cities is left to speculation. There is no doubt that, in 1966—as it became more and more apparent to a worried nation that rioting would continue—the Model Cities bill was the only fresh initiative Congress had to demonstrate its concern. But if Model Cities owed its survival in part to its identity as a violence antidote, the program's very design left no doubt that it was intended—like no other federal policy before it—to strike at the presumed roots of rioting. Yet in the first 9 months of 1967—a period during which Model Cities was being implemented by HUD—164 riots were recorded nationally. One observer notes, "It seemed at least possible that a full-scale national race war might break out" (Lemann, 1991, p. 190). It soon became obvious that mere adoption of Model Cities legislation had not been enough to calm the anger. The debate over the causes of the violence would become even more pronounced throughout 1967 and 1968.

It would be another year and a half after the Model Cities bill was signed before the presidentially appointed National Advisory Commission on Civil Disorders (the Kerner Commission) would issue its final report on the matter. Whatever the wisdom of its conclusions, the Kerner Commission report was preceded by considerable confusion about the reasons for the rioting that had occurred between 1964 and 1967. Yet certain realizations were taking root in America. First, riots most often erupted in run-down city neighborhoods inhabited mostly by poor and minority families. In that regard, urban rioting was more or less geographically and socioeconomically circumscribed. Second, although conflicts between civil authorities and citizens most often sparked the violence, many people

LIVERPOOL JOHN MOORES UNIVERSITY
LEARNING SERVICES

agreed that the root causes lay in the frustrations brought on by poverty, deprivation, discrimination, and lack of gainful enterprise.

Third, some people believed that a contributing factor was a sense of political powerlessness felt by those in the poor neighborhoods. With typically low voter turnouts and modest campaign contributions coming from poor areas, with city halls still dominated by white elected officials and bureaucrats, residents and businesses in these areas felt neglected and sometimes even actively exploited. Fourth, the many federal programs already in place to help these neighborhoods failed to grapple with the full dimensions of the problems plaguing their citizens. Even though Washington was spending billions of dollars on public and subsidized housing, Urban Renewal, food stamps, and health and welfare programs, for example, benefits were arriving piecemeal in inner cities. Even with a monthly federal welfare check, a poor family might still be living in a rat-infested slum tenement. Even with the benefit of periodic public health assistance, a son or daughter might drop out of high school or join a juvenile gang. Even where supposedly beneficial programs such as public housing and Urban Renewal were under way, families might be relocated to another neighborhood to make way for the vacant lots that followed the demolition of their old homes. In short, nothing Washington could offer at the time confronted the full range of tribulations plaguing poor and minority households.

If these were indeed the true causes of the rioting, then the Model Cities program was indisputably designed to respond to them. Locationally, its benefits were limited exclusively to central city *ghettos,* a term that quickly replaced *slums* and connoted to some observers images of Warsaw, oppression, and jack-booted authorities. The ghettos were portrayed as made up mostly of tightly packed poor and minority families; blight and deterioration were in advanced stages. Politically, Congress demanded significant participation by the disfranchised in these neighborhoods in the design and management of local Model Cities programs. With the exception of the CAP (enacted 2 years before Model Cities), nothing else in Washington's policy arsenal sought to defuse the frustrations of the politically dispossessed as forthrightly as Model Cities. Socially and economically, Model Cities recognized that the roots of discontent would not be severed by a panoply of single-purpose programs. A single program that could be made to address all or nearly all the frustrations of poverty and minority status was necessary. Model Cities allowed local people, including the poor, to decide for themselves what projects and services were needed. Moreover, it gave them the opportunity to design their own initiatives, as well as to patch together existing single-purpose programs, in a comprehensive response to the conditions fostering deprivation, discrimination, and rioting.

Just how intimately linked the Model Cities program was to the earlier years of urban violence is crisply articulated in a 1973 quote from a former secretary of HUD (almost certainly Robert Weaver). " 'I am sure,' " the former secretary notes, " 'that Model Cities would never have come out of the Congress if it were not for the riots' " (quoted in Button, 1978, p. 65).[2] Yet if riot responsiveness was a fundamental attribute of Model Cities, that identity carried with it a liability. For it was also easy to label Model Cities a program for rewarding the rioters. And some in Washington feared that President Johnson's new initiative would be perceived as confirmation that violence pays. For example, one Johnson administration official notes that, " 'every time there was a riot, that riot made it more difficult to get progressive social legislation passed' " (quoted in Button, 1978, p. 66). This official cites Model Cities as an example. The anger that exploded in Harlem and Rochester in 1964, in Watts in 1965, and in Cleveland, Chicago, and Philadelphia in 1966 brought beatings, shootings, stabbings, arson, and looting. In their wake, housing was destroyed, businesses were closed, jobs were lost, tax revenues disappeared, and municipal and state revenues originally meant for other purposes were diverted to repairing the damage. Some Model Cities opponents asked how could the federal government deplore such behavior—indeed, discourage it—while at the same time disbursing hundreds of millions of dollars in response to it? Does that not confirm the value of rioting?

As the Model Cities bill wound its way through Congress, administration officials and supporters were forced to walk the tightrope between the danger-of-violence and rewarding-the-rioters sides of the issue. Without measures such as Model Cities, some argued, more rioting would occur. No amount of official condemnation of violence would stop it. Only carefully designed programs addressing all the fundamental sources of discontent held any hope of success. In this regard, Model Cities could be viewed both as a program responsive to earlier riots and as a "preventive for future urban disorders" (Button, 1978, p. 73). It was enacted because "most federal officials felt [it] was the best possible answer at the time to the growing urban unrest" (Button, 1978, p. 66).

Model Cities supporters argued that the measure should not be identified publicly with the riots. It was not to be portrayed as a program to reward the rioters. Instead, Model Cities was intended to strike at the roots of poverty and despair. In the public discourse that followed enactment of the program, administration officials were careful not to bill it as a response to rioting. Instead, although Model Cities was criticized by Senator Robert F. Kennedy and others as "not enough," it was carefully coifed by White House, HUD, and congressional supporters to be the first critical step in a long march toward improving the quality of urban life for poor and minority citizens. Although only a 5-year experiment, Model Cities' results would lead, supporters argued, to the design of long-term and more effective programs to erase ghettos and alleviate the conditions of poverty. It was not a mere slapdash effort to reward rioters for their violence. Yet Model Cities remained the clearest expression of the United States government's effort to develop riot-responsive social legislation. This identity would become the very basis for dismantling the program once it became clear in the 1970s that rioting had subsided. More than a quarter of a century later, in the wake of the Los Angeles riots of 1992, the White House and Congress would grapple with this issue once again.

Tumultuous Times

With the flush of victory following passage of Model Cities, the Johnson administration could now point to a positive achievement in the face of criticism of Johnson's record on helping the urban poor from white liberals and black civil rights activists such as Whitney Young, Martin Luther King, Jr., and Roy Wilkins. But Model Cities was only one weapon in the arsenal of programs, regulations, and court decisions emerging during the mid- and late 1960s and early 1970s. Moreover, the tandem issues of civil rights and urban unrest were overshadowed in the minds of many Americans by the rising costs and casualties in Vietnam and the escalating protests and violence on college campuses. Antiwar activities embraced ever larger numbers of young people, the most radical of which were members of Students for a Democratic Society (SDS).[3]

Underlying public concerns such as these was the so-called sexual revolution. The effectiveness and widespread availability of birth control pills helped erode inhibitions among American men and women. Frank treatment of sexual issues in the mass media became common. The decline of single-sex colleges, the rise of coeducational dormitories, and the virtual abdication by many university administrators of parietal rules added to the ferment about traditional values and norms in America.

The rise of marijuana and other drug use among young adults was also troubling to many parents. The counterculture movement, symbolized by long-haired hippies and yippies, professed to reject orthodox American values. Some popular films conveyed a sense of futility mixed with a what-the-hell attitude about life in America. The center of gravity in popular music shifted from the relatively benign Beatles to the hard rock Rolling Stones. Acid rock animated a new musical argot among many American youth in the late 1960s. Even folksinger Bob Dylan's gentle acoustic guitar went high-decibel electric as the mood changed. "The times," as Dylan insisted, were "a-changing."

The new culture confounded and angered many older people. Even conventional political identities could no longer be relied on. As journalists struggled to distinguish what they called the New Left from the Old Left, readers grappled with where they themselves fell on the new political spectrum. Was Bobby Kennedy New or Old? If Eugene McCarthy was New, what had he been before he was New? If Lyndon Johnson and Hubert Humphrey were Old Left, what did that portend for the future of traditional liberals in the Democratic Party?

Only a thumbnail sketch is possible here, but the late 1960s was a confusing, unsettling, and divisive period in American history. No matter how profound, the violence in the cities was only one troubling issue among many competing for public attention. Urban rioting was clearly among the most threatening to the American status quo. Not only was the nation locked in a deadly struggle in Southeast Asia, it seemed to be headed for something akin to civil war within its own borders. Early on, small town and middle-class suburban families had been able to look on the violence in the cities as something like the Vietnam War (i.e., it was happening somewhere else). As rioting became more frequent, fears arose that the nation was headed for a full-scale race war.

Perhaps the most concrete threat for white Americans was the growing militancy of young blacks. Although civil rights groups such as the Southern Christian Leadership Conference (SCLC) struggled under King's leadership to pursue equality for blacks through peaceful means, other organizations were becoming impatient. The most visible of these was the Black Panthers. The Black Panthers was largely a national organization in name only, however. Its primary effect was in a few dozen cities such as Oakland, New York, and Philadelphia, where local chapters existed.[4] Not surprisingly, local law enforcement units kept a close watch on Panthers and were often quick to harass or arrest them—sometimes for good cause. Some Panthers and other black radicals called for open warfare against the police or whites in general. Others argued that integration was an unattainable—and perhaps undesirable—myth. Only separate communities, states, and even nations for black people offered the promise of freedom, Panthers insisted. Occasional gun battles occurred in some cities between law enforcement officers and Panthers or other black radicals. White America was frightened by these events, and many people saw them as further proof, if it was needed, that the nation was "going to hell."[5]

The vast majority of Americans wanted the rioting to stop, yet many whites, and nearly all blacks, felt that there was some justification for black frustration over inequalities. Of course, there were regional variations across the nation and antiriot and antiblack feelings appeared to be strongest in the South, in suburbs almost everywhere, in small towns, and among white ethnic neighborhoods in cities. These were the patterns to which mayors, governors, members of Congress, and the president often played.

Not to be overlooked is the influence that television, and to a lesser extent national opinion polls, had on those who participated in rioting. By the end of the Watts riots, there could be no doubt in the minds of young ghetto dwellers that violent action not only could produce ill-gotten goods, excitement, and "something to do," it could also make white Americans stand up and take notice of their plight. It could deliver jobs, housing, and other services through increased city, state, and federal assistance. For many young blacks, especially males, the family television set brought fresh evidence of the effects of rioting. Nothing else—no other action—got the attention of city hall, the state house, and Congress quite as effectively as mob behavior. Television coverage represented at the least a new catalyst to violence—one that had not existed during previous periods of urban mob violence in America. Civil disorders proliferated during the mid- and late 1960s partly in response to a heightened public consciousness about violence and its outcomes.

Newark and Detroit, 1967

In retrospect, there is a certain irony in the timing of the Model Cities legislation. The measure today appears to have slipped into existence during a lull between battles in the urban war. Riots in Chicago, Cleveland, and Philadelphia notwithstanding, deaths attributed to urban interracial mob violence dipped sharply in 1966. Although the number of cities experiencing these events continued to rise throughout the latter 1960s, 1966 brought a modest 9 riot-associated deaths. In 1965, there were 43 such tragedies (most in Los Angeles). In 1967, riot-related deaths leaped to 85 (Wikstrom, 1974, p. 22, Table 3). The worst riots broke out in Newark and Detroit.

By late summer 1967, it was apparent to everyone that whatever Washington had done to stem urban unrest, the measures had not taken effect. Seventy-one cities experienced at least one—sometimes two or three—outbreaks of mob unrest during that year. Moreover, the length of these confrontations increased; instead of 1-day events, more and more became 2- or 3-day cavalcades of looting, arson, assaults, and the like. The number of people arrested during 1967 rose sevenfold from the year before to more than 16,000. One study (Wikstrom, 1974) ranked 43 cities on the severity of their riots (i.e., the number of outbursts and their total duration in days). It found seven "major," 25 "serious," and 39 "minor" riots in 1967. Most had occurred during the steamy months of July and August, when tempers most easily flared. Among the most severe were riots in Cincinnati, Newark, Chicago, Wichita, San Francisco, Milwaukee, and New Haven, where total days of violence per city ranged from 5 to 7.

In Newark, the arrest of a black cab driver brought a firestorm of destruction beginning on July 12. Reported by police to be driving erratically, the driver, when stopped, allegedly struck one of the two arresting officers and cursed both. The driver claimed that he had committed no violations and that police beat him as he was being transported to the precinct station and once he was in a jail cell. The incident festered over the next few days, with black civic leaders and white authorities trading charges and denials. As tempers flared, the state police and the National Guard were called in. Fires were set in several stores and crowds began to loot. Sniping was widely reported. Police arrested more than 1,400 people; several blacks charged the police with brutality and verbal abuse. By early the following Monday morning, violence had finally subsided and 23 deaths had been recorded. Among them were those of a fire captain and a police detective ("The Newark Riot," 1971).

As shocking as the Newark riot was, the tragedy was soon overshadowed by an 8-day siege in Detroit that brought 43 deaths and more than 7,200 arrests (*Report of the National Advisory Commission on Civil Disorders,* 1968). On July 22, an early morning police raid on one of the city's many after-hours drinking clubs, known as "blind pigs," brought several arrests. Soon, a crowd of black citizens from the neighborhood gathered. As tensions rose throughout the morning, a rumor spread that police had bayoneted a black man. Rocks and bottles were thrown at officers and a few fires were set in stores. More than 300 state police were called in. By mid-Sunday afternoon, the number of fires was rising and looting had become widespread. Mayor Jerome Cavenaugh asked for National Guard intervention. He followed with a proclamation imposing a curfew from 9 o'clock at night to 5 o'clock in the morning. Later that evening, Michigan Republican Governor George Romney was flown over the city to survey the damage. A year and a half later, Romney would become President Nixon's secretary of Housing and Urban Development, perhaps due in part to his training in the Motor City that evening.

By 2 o'clock on Monday morning, more than 2,000 police and National Guardsmen were on duty in Detroit, with 8,000 more Guardsmen in transit to the city. As violence waxed and waned over the next several days, tanks and jeeps were brought in to help restore order. By late Tuesday, looting and firebombing had all but disappeared, thanks to the massive ranks of law enforcement personnel in the area. But open lawbreaking had been supplanted by covert actions such as sniper fire. Fires set earlier spread to nearby buildings, keeping firefighters on continuous duty. By Thursday, nearly all mob violence had ended and pressures were easing. By the following Tuesday, national guardsmen were removed and the curfew was ended. When the melee was finally over, almost 700 structures consumed by flames, about one third of these private homes (*Report of the National Advisory Commission on Civil Disorders,* 1968).

Although some of the 43 deaths were due to fallen power lines, mistaken identities, and accidental gun discharges, most happened when police or National Guardsmen shot looters, arsonists, or others suspected of or observed to be committing crimes. Merchants shot two people and rioters killed two others. Ten of the fatalities were white and 33 were black. As in nearly all such tragedies, not all the lawbreakers were black nor were all the good Samaritans white. Numerous incidents were reported of young blacks assisting firefighters and helping police officers direct traffic. Black community leaders fanned out into the streets of Detroit to calm tensions and deter the rumor mongers. In some cases, looters were white; in others, innocent blacks were arrested, injured, or shot by other whites or blacks (*Report of the National Advisory Commission on Civil Disorders,* 1968). When it was over, the catastrophe in the Motor City produced one half of all riot deaths occurring in 1967—as many as resulted nationally in the entire year of 1965 (*Report of the National Advisory Commission on Civil Disorders,* 1968; Wikstrom, 1974, p. 22, Table 3).[6]

The Kerner Commission

It was no surprise when Lyndon Johnson appointed his National Advisory Commission on Civil Disorders on July 29, 1967. Headed by Illinois Governor Otto Kerner, the commission was set up as the smoke from the Detroit riot was clearing. It demonstrated to the American public just how serious the federal government considered "the urban problem."

In a televised address to the nation on July 27, Johnson declared, "We have endured a week such as no Nation should live through: a time of violence and tragedy." He announced his new commission and, at the same time, warned that "looting, arson, plunder and pillage" would be punished and that the violence "must be stopped: quickly, finally and permanently." Calling for new "legislative action to improve the life in our cities," Johnson promised twice that there would be "no bonus or reward" for the perpetrators of crime in these incidents. He also pledged to "continue to search for evidence of conspiracy" through the FBI in rooting out the genesis of the disorders (*Report of the National Advisory Commission on Civil Disorders,* 1968, pp. 538-541). At the same time, he counseled compassion, understanding, and action toward the problems of the disadvantaged. His address adroitly walked a tightrope between a pledge not to reward rioting and an entreaty to Congress and local officials to help minorities achieve improved living conditions

At least eight episodes of urban unrest followed the Kerner Commission's appointment that summer, including serious riots in New Haven, Milwaukee, and Wichita (Wikstrom, 1974, p. 24, Table 5). Both East St. Louis and Chicago, the sites of similar violence a half-century before, suffered rekindled violence in September. Even as efforts to stem the unrest in the cities moved forward, resistance to further spending on urban programs rose steadily in the nation's suburbs and rural areas. Inevitably, the debate centered somewhere on the continuum between the danger-of-violence argument and the reward-the-rioters counterpoint. While resistance swelled among Republicans and conservative southern Democrats to War on Poverty initiatives, many moderate-to-liberal Democrats argued that the job was not complete and that other programs were necessary.

In March 1968, the much-heralded Kerner Commission report appeared on the newsstands in paperback form from Bantam Books, a commercial publisher. Because government reports were rarely viewed as marketable by publishers, this was a clear indication of the unusual level of public concern over urban violence. (Within 2 months the trade book was in its 12th printing.) The official Government Printing Office version of the Kerner Commission report would be released several weeks later. Running more than 600 pages, the Bantam tome contained a history of urban and racial violence in America, case studies and photographs of riots, sections on the causes and consequences of poverty and discrimination, and recommendations on courses of action. But what was remembered by millions of Americans, many of whom never read the study, was the single conclusive phrase in the introduction: "Our nation is moving toward two societies, one black, one white—separate and unequal" (*Report of the National Advisory Commission on Civil Disorders,* 1968, p. 1). Repeated endlessly by the mass media, it was both a warning of how serious the panel viewed conditions in urban America and an ultimatum on the consequences of inaction.

Discrimination and segregation, the introduction continued, "permeated much of American life" and "now threaten[ed] the future of every American" (*Report of the National Advisory Commission on Civil Disorders,* 1968, p. 2). If current conditions persisted, the report promised, American communities would become polarized and basic democratic values would be destroyed. It called for a national commitment to action, "massive and sustained" (p. 2) with "unprecedented levels of funding and performance" (p. 2). New attitudes, new expenditures, and, if necessary, new taxes were called for. White Americans, the report charged, had not fully understood, and black Americans could never forget, that white institutions created the ghetto, white institutions maintained it, and white society condoned it.

In another section of its report (*Report of the National Advisory Commission on Civil Disorders,* 1968, pp. 201-202), the commission reassured the nation that its staff had found no evidence, despite the president's admonition, of a conspiracy to foster rioting. It acknowledged that in some cases, individual leaders had deliberately sought to spur violence and radical action. But the commission found no basis for allegations or suspicions that any group had masterminded the rioting that had occurred in recent years.

Although public perceptions of the Kerner Commission report ranged from appreciation to antagonism, few could escape the underlying suggestion that an apocalypse lay just around the corner. Not since the Civil War a century before had the nation confronted violence and divisiveness on such a massive scale. Buried beneath the surface quotations, more fundamental choices were spelled out in the Kerner Commission report. It pointed out three policy directions open to the nation. One was a continuation of present policies with no significant change in funding or direction. The present policies option would keep the nation in low gear by drawing on the existing arsenal of programs such as Model Cities, Community Action, Urban Renewal, subsidized housing, health care, and remedial education. Alternatively, the report argued, Washington could shift into second gear with the enrichment choice. This, too, would continue current programs but would commit substantially greater funding to accelerate and further their effects.

Third, the integration choice would shift America into high gear. It would include enrichment and would add broad new measures by Congress to break down the barriers to full integration of minorities in American society. A centerpiece would be to open up the predominantly white suburbs to racial and socioeconomic integration, thus reducing the barriers blacks faced in finding decent housing, good schools, and safe streets. The study concluded that present policies was the least promising choice to end the conditions that contributed to rioting. The integration choice, although the most difficult, offered the most hope for bringing peace to the cities. Congress was then grappling with a new housing bill that would outlaw racial and ethnic discrimination in housing throughout the country. If this measure was enacted, it would put in place a critical stepping stone toward the integration choice.

The Kerner report singled out the Model Cities program as "potentially the most effective weapon in the federal arsenal for a long-term comprehensive attack on the problems of American cities" (*Report of the National Advisory Commission on Civil Disorders,* 1968, p. 479). The commission's compliment offered one more sign of the identity Model Cities had in Washington as a direct federal response to the conditions that led to urban rioting in America. Echoing its enrichment choice, the commission supported the president's request that Model Cities funding be raised to $1 billion for the next fiscal year and urged additional funding to target programs such as Urban Renewal. Many in the halls of the new HUD building hoped that these words would not fall on deaf ears.

Perhaps the ultimate paradox associated with the Kerner report was its timing. It was enough that the man who had created the commission, Lyndon Johnson, would stun the nation on national television at the end of the month by announcing his decision not to seek reelection to the White House in November. On top of this, only 6 weeks after the commission's report appeared on the newsstands, a crazed white racist, James Earl Ray, squeezed the trigger of his rifle in Memphis and took the life of the single most influential figure in the modern era of the civil rights movement. Even the nation's gaping disbelief in the wake of the Rodney King court verdict 24 years later could not compare with the outrage most blacks and many whites felt when Martin Luther King, Jr., awarded the Nobel Prize for Peace, was gunned down on April 4, 1968. That there was rioting in many cities was of little surprise, though of great consternation, to most observers. Predictably, seasoned riot cities such as Chicago saw outbreaks of violence in the aftermath of the assassination. Cities such as Pittsburgh and

Baltimore, relatively quiet until then, were suddenly paralyzed by arson, looting, and sniping. Perhaps most shocking was the nation's capital.

Washington had not suffered major racial violence since 1919. Long viewed as a relatively progressive haven for black opportunity, the District of Columbia had an appointed black mayor and a population more than two thirds black. With the King assassination, however, the nation's capital could no longer keep the lid on. With fires drifting an acrid haze over the city, sirens wailing constantly, and military vehicles patrolling the streets, there was ample proof that America's racial divide had widened. For Congress and the president, urban interracial mob violence was no longer confined to their television screens. It was now before their eyes.

Rioting after King's assassination and in the months that followed brought another dubious record of destruction to the nation. By year's end, 106 cities—about half again as many as in the previous year—had experienced 155 violent episodes of varying intensity. Almost 21,700 people had been arrested, nearly 2,800, had been wounded, and 75 had been killed (Wikstrom, 1974, p. 22, Table 3).

Civil Rights Act of 1968

The murder of the nation's foremost black leader and the ensuing riots played a part in convincing members of Congress to take a decisive step to combat racism. Within 1 week, Congress had embraced a major tenet of the Kerner Commission's integration choice by enacting the Civil Rights Act of 1968. Only a month earlier, an attempt in the Senate to filibuster the measure to death failed by only one vote. Signed by President Johnson on April 11, 1968, the law (often called the Open Housing or Fair Housing Act) put the nation on notice that the government would no longer tolerate discrimination in the rental or sale of housing based on race, color, religion, sex, or national origin. (Even as Johnson was signing the document, federal troops were patrolling sections of Washington's riot neighborhoods nearby.) Although racial exclusion from shops, stores, and public buildings was already a violation of federal law, the opportunity to purchase or rent residential properties was often legally denied to minorities. Few local or state laws barring such behavior existed; of these, fewer still were enforced. Consequently, there was little to prevent landlords or real estate agents from denying African Americans or other minorities the same opportunities available to whites.

Nor was it difficult for real estate agents to steer blacks, Puerto Ricans, and others away from white neighborhoods toward minority enclaves. Banks, through a variety of policies, frequently refused to lend minorities mortgage money. Mortgage insurance companies could deny a policy to blacks, making it virtually impossible for them to qualify for a mortgage loan. Appraisers had a professional code that advised that houses located in minority or racially changing neighborhoods should be appraised at a lower value than similar houses in white neighborhoods. The result of all these practices was patently clear. Although all minorities, including Jews, had felt the sting of racial or ethnic exclusion, blacks were disproportionately hurt. Largely confined to old inner-city neighborhoods, usually within a mile or two of the central business district, blacks occupied what were often the worst slums and tenements in urban America. Although low incomes and the cost of suitable housing had much to do with black housing choices, the devastating effects of discrimination were inarguable. Title VIII of the 1968 Civil Rights Act was a congressional milestone in the process of opening up America's suburbs to blacks and other minorities.[7]

Nevertheless, even as the list of progressive government programs and laws grew longer, tensions between blacks and whites continued to grow. A University of Michigan survey in the summer of 1969 found that 65% of American men were more worried about violent civil disturbances than any other condition in American life (Silver, 1968). Other surveys found that almost one half of whites believed "outside agitators" or Communists were among the perpetrators of urban riots in the mid-1960s. Less than one half felt that poor housing was a factor in these outbursts; even smaller percentages pointed to police brutality, white racism, or an insufficient supply of jobs (Silver, 1968).

On more traditional measures of white racial attitudes, though, there were signs that progress was being made. A Harris Poll (1971) found that in virtually every question posed, whites showed diminished racism when results were compared to those of a survey in 1963. Anywhere from one fifth to more than one half of whites had such perceptions as "blacks are inferior to whites," "blacks have less native intelligence," and "blacks have lower morals," however. A Gallup Poll (1972) found that southern white parents were much less likely in 1970 than in 1963 to object to sending their children to schools where blacks were enrolled. A similar trend was found among Northern parents, although the changes were not as dramatic.

Federal Response to Urban Poverty
and Mob Violence, 1960s Style

Even before Model Cities was enacted in late 1966, HUD had begun receiving application materials from many cities throughout the nation. As a result, in the days following the president's signing of the bill, HUD was able to announce the first 63 cities to receive Model Cities programs. Names of an additional 12 cities were released in March 1967, bringing the first-round total to 75. By November 1968, 2 years after the law was enacted, a second round of 72 cities had been awarded grants. The Model Cities demonstration would encompass 150 model neighborhoods in 147 cities (New York received three)—more than twice the number originally envisioned by the Wood task force.

It soon became apparent that, despite its precedent-setting identity, Model Cities was no more immune to politics as usual than other federal grant-in-aid programs. Former White House assistant Vaughn Davis Bornet (1983) points out that of the first 49 Model Cities awards, 48 went to cities in districts whose representatives had voted for the bill. Municipalities and a few counties in 45 states, plus the District of Columbia and rto Rico, were grant recipients. California communities received 11 gnations and New York received 10. Two states—Massachusetts and nsylvania—each received nine; Ohio, Michigan, New Jersey, and s (President Johnson's home state) received eight apiece. Perhaps se none of their communities chose to apply, five states—Missis-Nebraska, Nevada, South Dakota, and West Virginia—had no recipi-Geographically, 52 (35%) of the model neighborhood designations to communities in eastern states, 32 (21%) went to Midwestern states, ne remainder, 66 (44%), went to southern or western states (roughly speaking, the Sunbelt region).

As the United States struggled with its worst year of urban unrest, city administrations that had received HUD's blessing grappled with their Model Cities programs. Each city had to set up a CDA to run the effort and each CDA was to consult a Model Neighborhood Board. The board, made up of city residents, mostly those living or working in the neighborhood, met periodically and its members gave their advice and opinions. There was little surprise when many boards came under fire for not being representative of people in the neighborhood. Most boards were appointed; almost invariably, critics found fault with their composition. Even where boards were elected, however, voter turnout at these affairs tended to be low. Thus, boards were criticized for having too many people of one race, too few of another, too many homeowners and not enough renters, too many from certain sections of the neighborhood and not enough from others, too many businesspeople and too few residents, and so on. Concerns about representativeness plagued model neighborhoods just as they would confound Rebuild L.A. a quarter century later.

Tension also came about between the boards and the CDA staffs. Although many staff people were hired from the model neighborhoods, many more were from elsewhere in the city. Others were recruited from other parts of the country. Largely lower to middle class, CDA staff were sometimes suspected of careerism by some board members. In other cases, local mayors or city council members were resentful of the budgets controlled by CDAs and attacked staff or board members for ignoring traditional ward- and precinct-based patterns of protocol. Other problems were described in cities such as New Haven, where city hall officials, bypassed by CDA staff earlier, turned a cold shoulder to later entreaties for aid (Powledge, 1970). In a small Southern town, a cumbersome participation structure was set up. Seven planning subcommittees, a technical review committee, and a citizen's demonstration committee were part of the CDA effort. Even at that, the newcomer CDA director and his assistant spent months overcoming mistrust from many in the town (Rubin, 1981). In Atlanta, 66 separate committees were established to advise the Model Cities program on its plan (Marshall Kaplan, Gans, and Kahn, 1970b). The elected Boston Model Neighborhood Board suffered from poor attendance, a problem common nationally.

By 1970, most cities with Model Cities programs had completed their plans and were involved in setting up programs to build housing and other facilities and to deliver services in the neighborhood. Some localities placed emphasis on physical renewal, acquiring derelict buildings and renovating them or tearing them down. New units for low- and moderate-income families replaced the slums. Community centers, recreation or day care facilities, or parks were put in place. Counseling was offered to help residents improve their homes or buy and renovate houses in the neighborhood. Some communities stressed services such as job training, mental health, pediatric services, homemaking, drug addiction control, preschool and remedial education, citizen crime patrols, and the like.

The Boston plan, for example, contained 20 separate activities and projects (Haar, 1975). The Woodlawn Organization of Chicago proposed a dizzying array of "programmatic components" for its model neighborhood, numbering at least 30 projects and activities (Woodlawn Organization, 1970). In some cases, the smaller the community, the larger the aspirations. Gainesville, Georgia, a city of 18,000 people, put forth a plan for no fewer than 66 initiatives during its first action year (Rubin, 1981).

As the Model Cities experiment moved forward, it was clear that 150 model neighborhoods would come up with well over 1,000 individual projects and human service programs. In most cases, no two of these were designed exactly the same. Although this fostered enormous variety and individuality, it also made HUD's job of evaluation untenable. For if Model Cities were to emerge as a success, HUD must be in a position to tell Congress which approaches to reducing poverty and improving neighborhood life worked and which ones did not. Congress had created Model Cities as a laboratory for urban innovation. Definitive answers were critical if the original concept for Model Cities was to bear fruit. Without such insights, there was no way that effective new long-term urban programs could be designed and offered to cities across the nation.

One analyst (Schussheim, 1974) observes that about 20% of Model Cities funds were spent on education programs, 17% on environmental protection and development (including urban renewal), 16% on housing, 11% for health, and the remainder for concerns such as employment training and development, social services, crime prevention, recreation and cultural projects and economic development.

Nonetheless, HUD regional and Washington staff grappled with the unending complexity of Model Cities plans and programs. It was clear to them that, even if a program for, say, lowering school dropout rates, worked successfully in Miami, they could not make a convincing case that the same approach would work well in Dubuque or Peoria. Termed *the problem of uncontrolled treatment variation* by Rossi (1978), this dilemma grew out of the fact that there were simply too many differences between Model

Cities populations, school curricula, grading and promotion standards, and the like.[8]

As the Model Cities demonstration played itself out during the late 1960s and early 1970s, other dilemmas surfaced. There was constant bickering between CDA staff and model neighborhood boards in some cities and divisiveness and mistrust between city halls and these groups in others. Elected officials sometimes suspected board members of priming themselves for a run at public office (Schussheim, 1974). Even where relations were reasonably successful among these groups, the complex mechanics of day-to-day coordination could undermine progress. On top of these issues, mayors and city councilors were uncomfortable placing so much Model Cities program money in one section of the city when other poor and working class neighborhoods needed assistance.

In Washington and in regional federal offices, enormous amounts of time were spent in coordination. Virtually all 150 of the local Model Cities programs drew not only on the Model Cities supplemental grants for funding, but also on a host of narrower programs run by other HUD divisions and other cabinet agencies. Supplemental funds were sometimes called "rubber money" among HUD operatives because there was enormous latitude in their use. But the bureaucratic keepers of the many categorical grant programs necessary to carry out local Model Cities goals often resisted being coordinated. Countless mind-numbing meetings were convened in HUD's new office building near the Federal Mall in Washington. Shaped like back-to-back crescents, the unorthodox structure was itself emblematic of HUD's outsider status among venerable federal agencies.

Emissaries from departments such as labor, commerce, justice, and health, education, and welfare routinely banged heads with Model Cities executives. These agencies and many of their programs had been in existence far longer than either HUD or Model Cities. They reminded their hosts that they were forced to march to different congressional and regulatory drummers than the unique Model Cities program. The uses to which their money could be spent were often circumscribed by precedent, protocol, and legal interpretation (Schussheim, 1974). Aside from the Urban Renewal administration, none of the participating agencies had been awarded extra appropriations to feed Model Neighborhoods. HUD's new program meant that they were being asked to do more with not enough. According to Schussheim (1974), the original hope in HUD was that each dollar of Model Cities funds would generate three to four dollars from other sources. With the exception of the Department of Health, Education and Welfare, however, none of the other agencies ever came close to this benchmark. As it turned out, about one dollar of non-Model Cities money was spent for each Model Cities dollar during the first 2 action years.

In practical terms, this meant that a stalemate often developed. A Department of Labor (DOL) official, for example, might suggest a small DOL employment training grant for say, Seattle's model neighborhood. The rest of the necessary funds, he or she would argue, could come from the ever-accommodating pot of rubber money (i.e., Model Cities supplemental funds). HUD Model Cities staffers were constantly fending off these assaults on their limited budget. They learned that one of the detriments to having an enormously flexible funding source is that one could rarely argue that a proposed use was not covered under the Model Cities legislation. As a result, these meetings brought forth what could be called the Model Cities minuet. They resembled nothing so much as an elaborate and courtly dance, with each bureaucratic performer ritualistically circling strategically around the others, gesturing and bowing, maneuvering for position.

Getting other agencies to pony up was only part of the interdepartmental follies in which HUD participated. Another problem was contending with different programmatic emphases and mandates. For example, as a staff intern at HUD headquarters, I observed an interagency review of the Model Cities program in a moderate-sized Georgia city. The OEO representative was dragging his heels, withholding his office's support for its portion of the program. At issue was a plan to build housing for poor residents in the Model Neighborhood, hardly an inconsistent goal given the Model Cities legislation. On its surface, the project could not be more compatible with OEO's social welfare objectives. Nevertheless, the OEO representative was uncomfortable with the fact that the housing project was to be built on the other side of a railroad track considerably removed from the area where many whites lived. To make matters worse, it was to be located near an existing dump. From OEO's point of view, the project would perpetuate segregated housing patterns, possibly violating the Fair Housing Act passed by Congress a year earlier. In addition, placing the project on the other side of the tracks next to a dump conveyed a sense that blacks were not worthy of the same neighborhood conditions as whites.

But as far as Model Cities staff were concerned, the identity of their program as a segregation buster was precisely what had almost scuttled the legislation in Congress 3 years earlier. The dire predictions of Bronx Congressman Paul Fino and other conservatives were still ringing in their ears. Just as important, Model Cities staff, although idealistic by Washington standards (many were in their 20s and early 30s), did not have the messianic zeal of many staff from OEO, which was born in the heady aftermath of the New Frontier. Model Cities staff were ever mindful that President Johnson did not want to alienate city halls by circumventing existing elective power structures, as President Kennedy's OEO Community Action Agencies had. In the end, the Model Cities people argued that in this particular Deep South enclave, any new housing for poor people represented important social progress, even if not located among white families. They countered the OEO position by claiming that holding back approval of the plan in the hopes of relocating the housing to the other side of the tracks might sidetrack the project altogether. As it turned out, with time and many meetings, the Model Cities view prevailed. Yet such disjunctures were common as HUD struggled to bring other agencies into line.

Notes

1. One of the earliest discussions of the place-based character of Model Cities was put forth by Haar (1975). A high-level official in the Johnson administration in support of Model Cities, Haar later reflected on the program's tendency to constrict political support for itself by limiting beneficiaries to poor urban neighborhoods.

2. Johnson advisor Walt Rostow also shared this perception, leading Button (1978) to conclude that Model Cities was "spawned by the initial black violence" (p. 65). No new program, he adds, "was affected as much in both design and passage, by the early riots as Model Cities" (p. 66). Another study (Dommel, 1985), examining the evolution of federal community development policy, observes that Model Cities was enacted "partly to respond to the urban riots" (p. 477). Other observers have reached similar conclusions (Feagin & Hahn, 1973).

3. With chapters on hundreds of campuses, SDS became the symbol for a new brand of far-Left extremism. SDS had strong roots in its opposition to the Vietnam War and to what its members described as the oppression of blacks and other minorities. Yet the group never embraced more than a tiny percentage of the total student body on any campus. Although SDS and other campus radical organizations were occasionally involved in setting fires and planting bombs in campus buildings, the vast majority of members were inflammatory in word, dress, and manner only.

4. Wearing their hair in bushy afros and donning black turtleneck shirts and berets, members of the Panthers openly challenged white authority, sometimes through unlawful acts but more often through brazen postures. They often established support in black neighborhoods by running services such as soup kitchens and after-school recreation programs. On some occasions, though, they displayed weapons (albeit largely for the mass media), made gestures, and uttered profanities at whites in contemptuous tones.

5. Even at their height, the Black Panthers never enjoyed widespread support among blacks, much less whites. A Gallup Poll (1970) found that 19% of blacks surveyed gave the group a highly favorable rating and 36% gave it a highly unfavorable rating. Conversely, 83% of blacks gave the National Association for the Advancement of Colored People (NAACP) a highly favorable rating and only 1% gave it a highly unfavorable rating. Eighteen percent of whites rated the NAACP as highly favorable and 1% rated the Panthers as such.

6. For a quarter century, Detroit's outburst ranked as the single worst urban riot of its type in American history. It would take the jury acquittal of four police officers in the Rodney King court case to catapult Los Angeles into the dubious position of first place. Detroit and Los Angeles have traded this record back and forth since Detroit's first major riot in 1943 (the same year in which the infamous, though far less deadly, zoot suit riots befell Los Angeles). Like a major league pennant race, Motown's record fatality rate (34) in that year was tied by the Watts riot 22 years later. Quickly reclaiming the title of most deadly riot in 1967, the Motor City held on until 58 fatalities in South-Central Los Angeles brought it back to the City of ✱ Angels in 1992. Why such terrible tragedies have alternated between these two cities remains a mystery.

7. The law immediately prohibited discrimination in federally owned and privately owned, federally insured multifamily housing. By December 31, 1968, it forbade discrimination in all multifamily buildings except those of four or fewer dwelling units, one of which was occupied by the owner. By January 1, 1970, all single-family homes sold or rented through real estate brokers were affected ("President Signs Civil Rights Bill," 1968).

8. Uncontrolled treatment variation plagued HUD Model Cities staff as they struggled to compare "apples and oranges." Even worse for HUD, it could not hope to find the staff, time, and budget to evaluate hundreds of individual methods for dealing with problems such as teen pregnancies, drug addiction, unemployment, malnutrition, and recidivism among young offenders. HUD evaluation bureaucrats realized that, even if, for example, apartment abandonment rates by landlords dipped favorably in one model neighborhood, it might be difficult to link that outcome persuasively to a program run by the local CDA. Landlords react to many conditions when making such decisions, including mortgage interest rates, property tax burdens, the availability of rent subsidies, the stringency of building code enforcement, the demand for rental apartments, and the supply of apartments in the neighborhood (i.e., the competition). In short, HUD analysts would have to control for or weed out influences other than those included in the program itself (Rossi, 1978; Wood, 1990).

6

Model Cities Program Plays Out, 1969 Through 1975

W hen President Lyndon Johnson declined to run for reelection in 1968, Senator Eugene McCarthy (D-MN) was already vying for the Democratic Party nomination. But Johnson's decision opened the door to New York Senator Robert Kennedy's long-expected bid; Kennedy's entry eroded McCarthy's support. In the wake of Kennedy's assassination in Los Angeles in June 1968, the Democratic convention nominated Johnson's vice president, Hubert H. Humphrey, to head the party ticket. Many in the McCarthy and Kennedy camps, particularly the former's youthful idealists, were devastated by Humphrey's nomination. The sheer anger and futility felt by those opposed to the Vietnam War was only amplified by the tumultuous demonstrations, rioting, and police overreaction in Chicago that occurred during the Democratic Party convention in July 1968. McCarthy would live to run another year, whereas Kennedy, the last best hope of those who longed for a return to the New Frontier era of John F. Kennedy, was tragically lost to the nation. Although some Democrats viewed Robert Kennedy as overly ambitious, even ruthless at times, few doubted the man's courage or his commitment to civil rights. And no one doubted the voter appeal of the Kennedy identity. Now, however, not one, but two Kennedy brothers had lost their lives to assassins, and the family name rose to unparalleled heights of martyrdom. As a result, no one—particularly no one associated as closely with the Johnson administration as Johnson's vice president—could easily measure up.

Yet the indomitable Hubert H. Humphrey succeeded in winning his party's nomination as the favorite of party regulars. Richard M. Nixon, Eisenhower's vice president, emerged from political retreat in California to capture the Republican nomination. Alabama Governor George Wallace ran as an independent, largely on an anti-civil rights platform. By this time, lifelong liberals such as Humphrey were increasingly vulnerable to opponents such as Nixon and Wallace. The American voter was becoming increasingly weary of violence, the threat of violence, radical ideas, and be-trampled customs and mores. Yet many were also aware that under 8 years of Democratic leadership, the nation had made greater strides toward reducing racial injustice than at any period since the Civil War. Perhaps it was the abortive Democratic convention in Chicago that summer that turned the tide in the Republicans' favor. TV images of police riots against young—and sometimes violent—protesters and a raging, out-of-control Mayor Richard Daley conveyed an unsavory image to some voters. Nixon hammered away at the vulnerable voters' insecurities: the fear of class and race war, the continued rise of lawlessness, increasing federal costs to pay for urban damage and new programs, the escalating war in Vietnam. When it was all over in early November, Nixon's strategy had worked. Even with the more conservative Wallace taking 13.5% of the vote, Nixon edged out Humphrey 43.4% to 42.7%. A new day was about to dawn in America.

As Nixon's inauguration approached in January 1969, it was evident to some observers that the War on Poverty would succumb to the will of the silent majority. Whatever Nixon's ideological leanings regarding cities and the poor in America, he was first and foremost a pragmatist rather than an ideologue. From his perspective, the War on Poverty was losing its mandate. The great majority of American voters never had much tolerance for rioting in the cities; there were growing signs that even the more liberal wing of the Democratic Party was having trouble sustaining momentum for Great Society reforms. It was obvious that the billions in federal money committed to social action in the summer of 1968 had done little to allay discontent among black, civil rights, and social welfare organizations across the country. Moreover, with Martin Luther King, Jr. lost forever, there was no one who could continue the momentum of the civil rights movement with comparable dynamism, prestige, and charisma. On top of this, the staggering costs of the Vietnam War placed growing pressures on the federal budget. Programs such as Model Cities, Urban Renewal, public housing, and Community Action had consumed billions of dollars. Others, such as welfare, food stamps, aid to education, and public health programs spent billions more. Doubting the nation's long-term commitment to urban policy issues, Nixon recognized that Congress would look for vulnerable programs to cut to shift funds to the war effort and other domestic issues.

Another issue contributed to the Nixon administration's strategy. Although Democrats dominated America's city halls, Republicans were generally more successful in occupying governors' mansions. Nixon saw that it was to his party's advantage to reduce the influence of big-city mayors in Washington and to strengthen the hand of state chief executives. In practical terms, this meant that organizations such as the National League of Cities and the U.S. Conference of Mayors would lose their grip on Congress and the White House, whereas others, such as the National Governors' Conference, would enjoy a new legitimacy along Pennsylvania Avenue. Finally, Nixon knew that chances for reelection to a second term in 1972 would be enhanced if his administration could put its own imprimatur on federal urban policy. He believed that he could not simply continue the welter of War on Poverty programs and claim credit for their successes nor accept blame for their failings. The voices of civil rights, antipoverty, and urban organizations and their congressional advocates, although waning, were still too strident in Washington for Nixon to turn his back on the concerns of minorities and cities.

Model Cities Attacked, Then Terminated

One of the first acts of the Nixon administration was to appoint Harvard professor Daniel Patrick Moynihan to head the president's newly created White House Urban Affairs Council. Moynihan's role was to assist the administration in redesigning Washington's presence in the American city. Moynihan wasted no time recommending elimination of existing programs. He sought new strategies such as a national income assistance plan for families. In other offices at 1600 Pennsylvania Avenue, however, people with names such as Ehrlichman, Haldeman, Dean, Colson, and Magruder were busy trying to dismantle some of the machinery of the War on Poverty. Several Johnson-era programs, including Model Cities, came under attack. Over the next 5 years, the Model Cities initiative would be threatened at least four times before it was finally euthanized. First, Nixon put forward a proposal to reduce the program's budget by $500 million, redirecting the money to assist southern schools in desegregation activities. Intense lobbying by mayors, governors, and members of Congress thwarted this scheme. Next, the president announced termination of Model Cities funding, with future revenues coming from a new block grant program that Nixon's staff was designing. The block grant approach borrowed from the Model Cities supplemental funds concept, permitting localities to receive federal transfer funds with few strings attached. But it added another feature: Block grants would be an annual entitlement rather than a competitive grant. Although many mayors saw virtue in block grants, most felt compelled to keep faith with local Model Cities constituencies. They also wanted to avoid changing horses in midstream, possibly interrupting the existing Model Cities flow altogether. Again, successful lobbying, primarily by mayors, circumvented the threat.

Third, White House staff decided that if they could not immediately supplant Model Cities with block grants, they could achieve a similar end by converting the program to a "demonstration within a demonstration." As a first step, they succeeded in convincing 20 Model Cities communities to participate in a new Planned Variations experiment. Among other features, Planned Variations would allow city halls to spend Model Cities funds in other poor neighborhoods as well as the model neighborhood for similar types of projects. Although appealing to some mayors, Planned Variations meant that scarce existing resources would be further diluted, making it even more difficult for model neighborhoods to bring about substantive improvements. The new initiative went forward in the summer of 1971.

Nixon's fourth and final assault on Model Cities followed his landslide reelection to the presidency in 1972. Soundly defeating Democratic challenger South Dakota Senator George McGovern (60.6% to 37.5%), Nixon demonstrated that his political mandate to loosen the grip of the Great

Society on the federal purse strings had only grown since 1969. Frustrated by his inability to persuade Congress to enact a block grant bill, Nixon exercised executive prerogative and unilaterally suspended funding to several Housing and Urban Development (HUD) programs, including Model Cities. Leman (1991) reports that Nixon, emboldened by his reelection victory, sent a note to aide John Ehrlichman telling him, "Model Cities—flush it" (p. 218). The result was further erosion of effort and greater uncertainty in the 147 cities with model neighborhoods. During the latter half of 1973 and early 1974, Nixon would become more preoccupied with the Watergate crisis and other matters. Shortly after he resigned the presidency in 1974, Congress enacted the Housing and Community Development Act, terminating Model Cities, Urban Renewal, and several other HUD programs. Henceforth, cities could use Community Development Block Grant monies to finance nearly all purposes served under predecessor programs. As a quid pro quo, Congress exacted a "hold harmless" provision that would allow all Model Cities grant recipients to receive enough in block grant funds to complete their programs over the next few years (Frieden & Kaplan, 1975).[1]

Model Cities in Retrospect—
How Did It Perform?

By almost every measure, Model Cities was one of the most profoundly unconventional domestic federal program enacted since the New Deal. Unlike traditional single-purpose urban and poverty programs, it was a diverse multipurpose endeavor, interweaving many federal agency grants and loans into a separate custom-designed plan for each participating model neighborhood. Federal funds went not solely to city halls but rather to special organizations in which local citizen volunteers could share power with elected and appointed officials. Whereas conventional federal programs were rarely coordinated to address the needs of a single poor community, Model Cities sought to combine these programs under a single plan and focus them on a single needy area of each recipient city. Whereas earlier programs targeted physical and economic renewal or social services or law enforcement, Model Cities combined all these into a single attack on slums·and the needs of the poor. The highly flexible nature of Model Cities Supplemental Funds represented an unprecedented transfer of discretionary authority for program design between Washington and poor neighborhoods. Finally, prior to Model Cities, no other federal urban or poverty program had been enacted in direct response to a nationwide trend of urban domestic violence, nor had one ever been designed so implicitly to ameliorate the presumed causes of such violence. More so than any other single federal initiative, therefore, it was an antiriot program.

Of course, this does not imply that those who supported the Model Cities bill were merely riot busters. Some who voted for the Demonstration Cities and Metropolitan Development Act in 1966 were more interested in sections of the bill other than Model Cities. Some knew that the measure in its final form could not possibly change the ghettos all that much; at most, it might buy time and yield some useful insights. Model Cities might be perceived by city officials, civil rights groups, and others as a show of faith in their worthy causes. But it is safe to conclude that most who supported Model Cities sought, among other things, a cessation of the violence in America's cities.

What, then, came of Model Cities? With all the torturous accommodations necessary to enact the bill, with all the management headaches in the bureaucracy, with all the petty squabbles in 147 local governments, what did the nation learn about stemming the causes of rioting? It is tempting to point out that for whatever reasons, urban interracial mob violence gradually faded away after 1968. Few people, however, would be willing to imply a cause-and-effect link between the War on Poverty and the cessation of rioting. Virtually none of the conditions associated with economic deprivation, racial discrimination, and neighborhood deterioration had substantially changed by the early 1970s.

Many observers have reflected on Model Cities since the early 1970s. With painful unanimity, they make clear that the nation lost the opportunity to carry out its original plan. Never, in a truly systematic way, were the individual successes and failures, gains and losses, of hundreds of local efforts carefully evaluated. Were there any "winners" among the gaggle of city job training programs? Did any projects successfully encourage poor children to remain in school? What about drug abusers? Did any locality develop inroads into weaning young people away from heroin? Sadly, we will never know. For so overwhelming was the national momentum to get on with other matters during the Nixon, Ford, and Carter years that few in the White House or on Capitol Hill wanted to unearth the painful reminders of the War on Poverty. Furthermore, Nixon's efforts to disengage Washington officials and turn urban policy decisions over to state and local governments succeeded in convincing many that solutions to poverty, neighborhood decline, and discrimination would no longer be found entirely in Washington (Beauregard, 1993). Believing that, one could find little value in picking over the remains of Model Cities. If the Feds would no longer have a hand in crafting new initiatives based on the evaluations, why bother? If anything, it made more sense for evaluations to be carried out at the local level. After all, with Nixon's block grants, municipal governments would now have the resources to fashion their own solutions to local problems. Yet few, if any, such efforts came about.

Further complicating the evaluation effort was the change in presidential administrations following the 1968 elections. A more or less complete turnover in Schedule C (i.e., political or noncivil service) appointees at HUD introduced a new cast of characters in the upper echelons, headed by Secretary George Romney. A former governor of Michigan and witness to the Detroit riots a year and a half earlier, Romney, although not unsympathetic to the goals of Model Cities, was not overly impressed with the arguments for evaluation. His attention was directed to Operation Breakthrough, a HUD series of demonstration projects to showcase promising manufactured housing techniques. More in keeping with Republican philosophy about the role of the private sector in social progress, Operation Breakthrough and other HUD initiatives represented for the Nixon administration the Republican future rather than the Democratic past. With no mandate to continue Model Cities past its congressional deadline, Romney and his lieutenants found little incentive to champion an exacting analysis of the program's effects and accomplishments. To an administration dedicated to reducing the role of Washington in local affairs, what purpose could there be in spending money on efforts that were based on the assumption of continued federal involvement in these matters?

One authoritative study (Frieden & Kaplan, 1975) points out that, almost from the start, HUD evaluation staff faced an uphill battle in mounting an effective assessment of their program's effects on the quality of life for residents living in model neighborhoods. Little attention was paid to this important mission early in the program's life because, among other concerns, Model Cities personnel were overwhelmed with staffing, organizing, and reviewing applications, plans, and first-year action reports. Later, as Romney and his lieutenants took root at HUD, priorities shifted to reflect the Nixon administration's evolving directions.

Just as troubling were methodological issues. The difficulty of developing a consistent set of standards and procedures for evaluating what amounted to thousands of individual projects in 147 cities continued to plague Model Cities staff. On top of this, Model Cities staff charged with managing daily program operations and providing technical assistance were suspicious of, if not hostile to, those responsible for evaluating the successes and failures of their efforts. As time went on, the Model Cities administration placed more emphasis on compelling local city demonstration agencies (CDAs) to evaluate their own model neighborhood programs. But because of time and budget constraints and the fear of looking bad in the eyes of HUD staff, even these efforts bore little fruit.

Two book-length studies of Model Cities (Frieden & Kaplan, 1975; Haar, 1975) found very few evaluative studies commissioned by the federal government. Furthermore, because both books were published before the 1980 decennial census, timely, accurate, and comprehensive data were unavailable to analyze for evidence of possible program effects. Several articles and papers were also published, but only a few are able to provide insights linking Model Cities to the urban poor (Schussheim, 1974; Wood, 1990).[2]

Instead of careful evaluations of program outcomes, HUD confined itself largely to a series of studies of the planning and management processes within the Model Cities administration and in a small selection of local CDAs around the country. Although not unimportant, these analyses were concerned primarily with feeding HUD staff information about concerns such as coordination of procedures, timetables, and funding cycles. Some also documented difficulties in administration of local programs such as rivalries between CDAs and model neighborhood boards or between either or both of these bodies and city halls (Marshall Kaplan, Gans, and Kahn, 1969, 1970a, 1970b, 1973a, 1973b). The Brookings Institution examined issues of local governance and citizen involvement in Model Cities programs in 16 cities (Sundquist & Davis, 1969).

As far as HUD officials were concerned, the practicality of these reports was borne out by the feedback they provided to allow staffing, procedural, and policy changes to be made in HUD's Model Cities administration. They also allowed HUD executives to demonstrate to those who still cared on Capitol Hill that HUD was, indeed, following through on the original Model Cities concept of demonstration and evaluation. But evaluation, in these early cases, was little more than a pulse-checking exercise designed to keep HUD Washington staff in touch with events in the field. Tragically, no systematic government campaign ever followed these initial studies to assess the extent to which local programs ultimately achieved any significant and lasting changes in the lives of those living or working in model neighborhoods.

Nonetheless, a few small studies of local results were carried out by consultants and academic researchers. By the middle of 1972, the Model Cities administration had transferred a total of $1.7 billion to 147 cities; six localities received one half of this amount: Chicago, New York, Seattle, Newark, Boston, and Savannah. One research project (Washnis, 1974) looked at model neighborhoods in these cities. Although Washnis (1974) did not attempt to evaluate program outcomes, he offered some unpromising prognoses as cities completed their second and third action years. On overall program funding, for example, Washnis observed: "Model Cities never received the necessary funds to overturn urban blight. Nor have many of these cities been able to maintain what they started" (p. 7).

Four of the six cities were among the 20 communities that had agreed to participate in the Nixon Planned Variations experiment. This allowed them to spend Model Cities funds over a broader geographical area than was covered by the originally defined model neighborhood. Washnis (1974) noted that, in Seattle, Dayton, Indianapolis, and Newark, Model Cities monies were being shifted away from human services such as welfare and job training—needs most keenly felt by poor residents—to bricks-and-mortar projects such as housing and street improvements, which usually cater to property-owning, middle-class interests. Washnis concluded with a mixed picture:

> It is difficult to see any measurable improvement in the overall quality of life for entire Model Neighborhoods. For the most part, housing and streets are still in a deteriorated state, unemployment is high and survey after

survey shows citizens' discontent with their lot and a remarkable lack of understanding about the objectives of Model Cities. The elimination of outdoor toilets, the paving of a few streets and the construction of some scattered homes have not been enough to raise the spirits of the mass of people. Yet, in spite of this, there are improvements in some quality-of-life goals. (p. 21)

Another consultant (Booz, Allen, & Hamilton, 1971) was asked by HUD to examine what effect Model Cities was having on people's desire to move into or out of model neighborhoods. There had been some speculation at HUD that the new housing and services available would attract families to move into the model neighborhoods. Others suggested that some families, especially homeowners, would not like being associated with government welfare programs and might move away. Neither event was a desirable goal of the Model Cities program, so interest in the study was limited to reassurance that the program was not contributing to immigration or outmigration.

Consultants visited 13 model neighborhoods around the nation and interviewed more than 500 people during late 1970 and early 1971 (Booz et al., 1971). They found that in no city did the number of movers exceed 3% of the model neighborhood population. Slightly more than one half (55%) moved out of the neighborhood, whereas the remainder moved into it. This slight balance of outmigration amounted to less than 2% of the model neighborhood population in every case.

The study (Booz et al., 1971) also probed the flow of Model Cities program funds. It found that of the $99 million in supplemental and categorical funds allocated to the 13 cities, about one half had been spent by the time the study was underway. Approximately 60% of this "was spent outside the Model Neighborhood area to purchase goods and services in support of Model Cities activities" (p. iv). Like population migration, the dollar flow was modestly outward. The local CDAs were spending a low of $5 per resident and a high of $83 per resident. Clearly, Model Cities money was helping bankroll neighborhood economies by increasing the cash flow. Moreover, there were no signs of a substantial flow of funds out of those neighborhoods. Although purchasing more goods and services inside the neighborhoods was desirable, it was not a professed goal of the Model Cities program. HUD staff could take comfort in most of the study's findings.

Urban Conditions Index

Although Model Cities achievements remain largely undocumented, the contributions of many of the War on Poverty programs to urban life were at question during the late 1960s and 1970s. Among the results were efforts by some scholars to assess the overall contribution of such programs to the social and economic status of cities. Attempts were made to develop a quantitative system for gauging the relative condition of cities. Usually called an index, one of the most widely discussed was the *Urban Conditions Index* developed by Fossett and Nathan (1981). Examining 53 of the 57 largest cities in the nation, the authors devised an index composed of a single number for each city computed from a combination of the city's age (as measured by the amount of older housing), population loss, and per capita income in 1960. A second index was computed from corresponding data in the 1970 census. Standardized to a mean of 100, the index indicates that cities with scores above that number are worse off or "more distressed" than cities with scores below 100.

Fossett and Nathan's (1981) findings demonstrate that between 1960 and 1970, among the 53 largest cities, the Urban Conditions Index rose in 28 cities, indicating a worsening of conditions. Nearly all these were located in the Northeastern and Midwestern states (i.e., the Frostbelt). The remaining 25 cities (47%) experienced a decline in the index, a sign that conditions were improving by some measure. Most of the improving cities were located in the Sunbelt region of the nation, composed of the Southeastern and Southwestern United States, including California. Of the 53 study cities, 41 had Model Cities programs. Of these, 15 (36%) had a decreasing Urban Conditions Index during the 1970s (i.e., improving conditions), whereas the 26 remaining cities (64%) had an increasing or worsening index. Thus, whatever effects increasing federal urban aid was having during the 1960s, many cities—including Model Cities participants—had not yet begun to show signs of improvement, as measured by the index. I should acknowledge, however, that although several federal urban programs were underway by 1970, local model neighborhood programs were just beginning to be operationalized. Nevertheless, Fossett and Nathan carried part of their analysis into the 1970s, finding that 34 cities experienced continued population decline from 1970 to 1977. The remainder, 19 cities, experienced either no loss or a gain in population; all but one of these were cities with a 1960-1970 index below 100. Similarly, measures of income show that the number of cities with per capita incomes below the average for the 53 cities in the study rose from 25 to 28 between 1969 and 1974. The Fossett and Nathan study demonstrates that, given certain indicators, there was little justification for optimism by the mid-1970s insofar as the fallout from federal expenditures was concerned.

Urban Decline Index

A more expansive index was crafted by the Brookings Institution (Bradbury, Downs, & Small, 1982). This encompasses a larger group of cities and computes an index incorporating four, rather than three, measures of urban conditions. Moreover, it is expressed as a longitudinal, rather than static, measure, showing change over the first half of the 1970s. Termed an *Urban Decline Index,* it combines measures of the unemployment rate (1970 through 1975), violent crime rate (1970 through 1975), city government debt burden (1971 through 1975), and percentage change in per capita income (1969 through 1974). Thus, increases in unemployment, violent crime, and debt burden and decreases or slower rates of increase in per capita income are viewed as decline. Under each of the four indices, cities were ranked according to their degree of severity. Each of these four distributions was divided into thirds; those in the upper one third were assigned a score of +1, those in the bottom one third were assigned a −1, and the remainder were assigned a 0. A summary score was calculated for each city by adding the four scores it achieved. The most severely declining cities were rated a −4 (all four measures showed negative trends), whereas those with the least evidence of decline were rated +4 (all four showed positive trends). The authors examined 153 American cities of at least 100,000 population in 1970.

The Brookings study (Bradbury et al., 1982) found that the largest central cities, especially older industrial cities of the Northeast and Midwest, were most heavily represented among those with the worst decline rating (−1 to −4). Cities such as Boston, Cleveland, Dayton, Hartford, Jersey City, and Trenton had among the most severe unemployment, violent crime, city debt, and per capita income measures. They were closely followed by Atlanta, Detroit, New York, Philadelphia, Rochester, and Worcester, among other cities (–3). Yet the presence of Las Vegas, Nevada, and Riverside, California, demonstrate that even Sunbelt cities are not entirely exempt from the worst urban conditions. The older Frost-belt cities were most heavily represented among cities scoring –2, although several California cities and Tucson were also included. Among the 57 cities scoring –1 to –4, 14 (25%) are located in the nation's Sunbelt (many in California or Florida). As decline scores rose into positive numbers, the share of non-Frostbelt cities rose as well.

Table 6.1 compares Urban Decline Index scores for all cities in the Brookings study (Bradbury et al., 1982) with a subset of those cities— those that received a Model Cities designation. Fifty-four percent of the 153 study cities were Model Cities designees. The data show that almost one half (48.8%) of cities receiving Model Cities awards scored in the decline range (–1 to –4), whereas a little more than one third (37.3%) of all cities did so. These data demonstrate that, generally, when Model Cities programs were in peak operation during the first half of the 1970s, the combined effects of unemployment, violent crime, municipal debt, and per capita income lag were worse in the Model Cities communities than in cities overall.

Although not definitive in their findings, these indices offer little confidence that the War on Poverty's effect—at least over the short run— was having measurable positive effects on America's older cities. Yet it was precisely at this time (during the early 1970s) when programs such as Model Cities were under greatest pressure to demonstrate their effective-ness. Although urban constituencies were still well organized and visible, public concern about the poor and cities was still in the air, and further urban interracial mob violence was still viewed as a threat, the opportunity remained to strengthen urban policy. Significant progress in reducing poverty, unemployment, crime, and other measures of urban quality of life would have to be demonstrated to Congress. Most urban analysts would argue that public policies grappling with the exceedingly complex dilem-mas of the urban poor could not and should not be expected to reverse these dynamics in so short a period of time. How can we expect to see measurable progress in reducing poverty and urban decline after only a decade of operation? Yet more recent studies of programs such as Job Corps and Head Start have found evidence of just such promising progress (Levitan & Johnson, 1986). Unfortunately, as the Nixon-Ford era demonstrated, the

TABLE 6.1 Urban Decline Index, 1969-1975

| | All Cities (N = 153) | | Model Cities Only (N = 82) | |
Index Score	No. of Cities	Percentage of Cities	No. of Cities	Percentage of Cities
−4	9	5.9	8	9.8
−3	10	6.5	11	13.4
−2	20	13.1	9	11.0
−1	18	11.8	12	14.6
0	29	18.6	15	18.3
+1	30	19.6	14	17.1
+2	23	15.0	7	8.5
+3	10	6.5	5	6.1
+4	4	2.6	1	1.2
Totals	153	99.6	82	100.0

SOURCE: Bradbury, Downs, and Small (1982, pp. 51, 56). Reprinted by permission of the Brookings Institution.

political attention span would not persist for so lengthy a period of time and the nation would move on to other issues and confront other challenges. Although urban interracial mob violence would largely dissipate over the 1970s, it would not disappear altogether.

Paradoxically, there are signs that Model Cities and other Great Society programs may have had their most profound effect in areas never expressly intended. A study carried out in the early 1980s in 10 northern California cities sought insights on how programs such as Model Cities had organized and acclimated blacks and liberals into coalitions to support progressive social change (Browning, Marshall, & Tabb, 1984). The focus was not on the effects of the Model Cities program on residents of model neighborhoods, but rather on a secondary outcome—political empowerment. The authors found that the stronger the liberal minority coalition formed under the local Model Cities program, the more effective it was in minimizing funding cuts when the Community Development Block Grant program replaced Model Cities during the mid- and late 1970s. The authors conclude that Model Cities and similar federal social initiatives helped acculturate blacks and other minorities to the necessity to mobilize in order to achieve their objectives. Where this was so, city governments, in turn, were more responsive to their interests. There was, as Schussheim (1974) points out, a "tendency to disestablish the old-line white liberal leadership of some organizations and replace them with younger, more group-conscious minority directors and boards" (p. 131).

The Mystery of Model Cities

Of all place-based federal programs designed to alleviate the conditions of poverty, Model Cities was the most comprehensive in its intentions. Surely, it was one of a very few (if not the only) created to respond directly to the conditions contributing to mob violence. Its profound departures from convention notwithstanding, Model Cities to this day remains a mystery to those who designed it, supported its enactment, administered it, or attempted to analyze its performance. Neither Congress nor the American public knows what effects, if any, Lyndon Johnson's most unusual domestic policy initiative had on poor neighborhoods. More than a quarter century later, our cities remain plagued by crime-ridden poor neighborhoods in which drugs alleviate despair and welfare is a surrogate for well-being. Even though the original catalyst for passage of Model Cities—the epidemic of urban rioting—has diminished in frequency, the rioting in South-Central Los Angeles showed us that the capacity for megaviolence persists. Although we can only speculate about the effectiveness of Model Cities in reducing poverty and the conditions that contribute to mob violence, we cannot conclude at this point that its positive effects were very substantial or enduring. Some observers believe that if the program had received continued support from the Nixon administration, its true reach would have been demonstrated (Wood, 1990). I argue later in this book, however, that even if place-based initiatives such as Model Cities were potentially efficacious for their times, changes in social, economic, and political trends in the intervening years have diminished their promise.

In the next chapter, I explore the course of urban interracial mob violence over the quarter century since the Great Society ground to a halt. I also examine the response by the White House and Congress to the catastrophic aftermath in which the nation launched a new, albeit far less grandiose, strategy to address some of the causes of urban interracial mob violence.

Notes

1. In an ironic twist, Nixon's long-sought campaign to place his imprint on federal urban policy finally bore fruit—only it was Gerald R. Ford, not Nixon, who signed the new act. And it was during the Ford administration that the federal Model Cities program—America's most comprehensive effort to reduce the causes and conditions contributing to urban rioting—met its demise.

2. In addition, a few largely descriptive case studies of local programs have been published. For example, one case study of the Model Cities program in San Antonio, written at the request of Congressman Henry B. Gonzalez (D-TX), gives a positive review of the results there (Woods, 1982). Another study examined the local program in Cohoes, New York, offering a basically supportive view of Model Cities (VanBuskirk, 1972). A study of Gainesville, Georgia's, program confined itself to organizational and political matters; it quotes one local Model Cities official who felt that the program was "designed to fail," noting that public expectations exceeded both its potential and its performance (Rubin, 1981, p. 24).

7

From Reverend King
to Rodney King

The years during which the Nixon administration prevailed in the White House were not unmarked by urban interracial mob violence. By comparison to the Great Society years, however, an uneasy calm befell the nation's cities from 1969 on. Moreover, by January 1973, the United States had found peace abroad, negotiating an end to the involvement of American troops in Vietnam. New issues such as the women's movement, gay rights, and environmental protection drew momentum from the civil rights and urban crusades of the previous decade. The devastating Arab oil embargo and severe petroleum shortages threw the nation into anxiety and were followed by the economic recession of the mid-1970s. As if these issues were not enough, New York and other cities suffered severe fiscal crises threatening municipal bankruptcy. And of course, the Watergate affair vastly undermined public confidence in Washington and the federal establishment. As if to round out a decade of eroding American faith in the nation and its future, the painful and protracted ordeal of American hostages in

Iran showed that even one of the most powerful nations on earth could be "brought to its knees" by the actions of a relatively small country. Not even America's exuberant bicentennial celebration in 1976 could sustain the public's spirit for long.

If the United States was preoccupied with matters far removed from the day-to-day realities of life among the poor and minorities in its cities, the poor and minorities themselves would remain quiet for about a decade before urban interracial mob violence would explode again. A Democrat was once more occupying the White House when the disorders occurred. Jimmy Carter, former governor of Georgia, was in the sixth month of his term when a bizarre outbreak of looting and burning occurred during a massive power failure in New York City in July 1977. Taking advantage of the cover of darkness and police distraction, thousands of people plundered retail shops and rampaged through the streets, injuring more than 400 law enforcement officers. Unlike other riots, however, there appeared to be no central grievance precipitating crowd behavior. Yet more than 2,000 businesses were reported looted and about 1,000 fires were set. Should the matter be discounted as merely opportunistic lawlessness—in the words of Edward Banfield (1974), "rioting mainly for fun and profit" (p. 213)? No consensus ever evolved.

During his term, President Carter sought to revive and modify Washington's urban agenda. To lead the charge, he appointed as secretary of Housing and Urban Development (HUD) former Howard Law School Dean Patricia Harris, the first black female cabinet secretary in U.S. history. She redirected the Community Development Block Grant program to concentrate more of its funds on the needs of poor communities and low- and moderate-income households. A new Urban Development Action Grant program was enacted during Carter's term to spur public and private financing of downtown projects designed to create jobs and enhance tax bases. Carter's administration also introduced a proposal to establish a federal community development bank to assist community groups in neighborhood revitalization efforts. (The measure failed to survive congressional review.) Harris restructured HUD to place greater emphasis on neighborhood organizations and community development needs of minority and ethnic communities. In charge of this section was Father Geno Barone, an activist Catholic priest who in many respects symbolized the shift Carter sought from the War-on-Poverty identity of the Johnson administration. Associated primarily with urban white ethnic and blue collar working-class groups, Father Barone embodied Carter's hope that federal urban aid could embrace a broader spectrum than solely the minority and low-income populations associated with policies of the late 1960s. Thus, interests represented by Nixon's middle Americans and the silent majority found their way into Carter's expanded urban policy perspective. This, Carter believed, could broaden congressional and popular support for a revived commitment to the cities.

⁂ Miami, 1980

In Carter's last year in office, months before he would be defeated for reelection by Ronald Reagan, a massive riot devastated sections of the Liberty City area of Miami (Portes & Stepick, 1993). The disaster occurred in response to the acquittal by an all-white jury of four white police officers charged with beating a black businessman to death. The riot raged from May 17 to 19, 1980. Eighteen people were killed, more than 400 were injured, approximately 1,100 were arrested, and damage was estimated at about $200 million ("Property Damage Exceeds $100 Million," 1980). The tragedy had began 2 months previously, when a black insurance agent, Arthur McDuffie, was stopped by police for a traffic infraction. A chase ensued and the man was apprehended and later died from injuries. The arresting officers insisted that McDuffie's injuries were accidental and occurred during the pursuit. They were subsequently charged with contributing to his death and with concealing facts. Their trial was moved to Tampa in an attempt to find a jury untouched by the tensions between whites, blacks, and Hispanics in Miami. Within a few hours of the time the four officers were acquitted, violence commenced in the predominantly African American Liberty City section of Miami.[1]

At the time, it was the worst outburst of urban interracial mob violence since the widespread disorders following the assassination of Martin Luther King, Jr. The catastrophe brought President Carter to Miami on June 9. Perhaps sensing his political vulnerability in appearing to reward the rioters, Carter cautioned that help from Washington would be limited. Five months later, the president was roundly beaten by Reagan, 50.7% to 41%.

Miami, 1989

During the 1980s, President Reagan succeeded in vastly undermining the nation's support for cities and the poor. Calling on the Congress to cut taxes, federal spending, and government regulations, the Reagan administration argued that nothing could rekindle the American economy as much as less government. Reagan found a responsive chord in the American public. He was seeking nothing less than the dismantling of the remains of a half century of predominantly Democratic social and urban programs. A job initiated by Richard Nixon, this mission was considerably expanded under Reagan. His administration sought to reduce the federal budget by billions of dollars, proposing to eliminate urban domestic programs such as General Revenue Sharing, Community Development Block Grants, Urban Development Action Grants, and the Comprehensive Employment and Training Act. (Only the block grant program survived the attack.) In addition, Reagan attempted to privatize government-subsidized programs such as the Amtrak rail system and public housing. Such efforts, although generally unsuccessful, would have turned over responsibility for ownership and operation of the rail systems to private enterprise. Public housing units would be sold to their tenants.

In the end, Ronald Reagan and sympathetic members of Congress succeeded in reducing the HUD budget from $36 billion in 1980 to $18 billion in 1987. Assisted housing authorizations dropped by $20 billion and rental housing production dropped by 110,000 dwelling units. Although overall federal support for urban and domestic programs had begun to decline (in constant dollars) in the second half of Jimmy Carter's term, Ronald Reagan oversaw a far more drastic reduction (Advisory Commission on Intergovernmental Relations, 1992; Reischauer, 1986).

If Reagan's inner circle wanted to believe that 1960s-style violence was a thing of the past, they would first have to overlook the Miami riots of 2 years earlier. But that tragedy—like its predecessors in the late 1960s—had occurred while a Democrat was in the White House. In light of the comparatively peaceful years characterizing the Nixon-Ford era, it was not difficult for Reaganites to be lulled into a false sense of security. But this time, not even a conservative Republican in Washington could preside over a riotless term. Miami was the stage on which the violence was played out again. A few days after Christmas 1982, Miami police shot a youth in a game arcade in the predominantly African American Overtown section. A mob assembled and charged a liquor store, causing $50,000 in damage. Between December 28 and 31, 26 people were injured and 43 were arrested. Although this was a minor incident in comparison to the unrest of $2\frac{1}{2}$ years earlier, the Overtown event underscored the rising level of tensions between the growing Hispanic population in Miami and the largely poorer, less affluent black population. It also signaled to the nation that Miami simmered with frustration and anger and could boil over at any moment. Six years later, it did.

The city was to be the site of the 1989 Superbowl football championship and was in the midst of preparations. On January 16, a Colombian-born police officer shot an African American motorcyclist speeding from the scene of a suspected crime in Miami's Overtown section. A police officer, William Lozano, had allegedly ordered Clement Lloyd to stop. Lloyd, a police cruiser chasing him, had raced the motorcycle toward Lozano, who fired his weapon (Portes & Stepick, 1993). Lloyd was killed; his black passenger, Alan Blanchard, died when the vehicle was involved in a collision. The incident set loose a reign of looting, arson, beatings, and shootings. A police substation was attacked by a crowd and attempts were made to set it afire. Two whites traveling through black neighborhoods were injured. A 130-block section of Overtown and the adjoining Liberty City area were sealed off by police later that day. One person was killed, six were shot, 27 shops were torched, and approximately 400 suspects were arrested in the 3-day melee. Miami public officials and business persons wrung their hands in fear that Superbowl crowds would avoid the city that provided the setting for the television series *Miami Vice*.

Officer Lozano was arrested and charged with manslaughter (Hackett, 1989). He was found guilty and sentenced to 7 years imprisonment in his first trial, but the verdict was overturned by the Third Florida District Court of Appeals on June 25, 1991. The court ruled that the conviction may have resulted from the jury's fears of another riot should the officer be acquitted. The appeals court decision was upheld in May 1993 and the officer was acquitted by an interracial jury in Orlando. The trial had to be moved five times before a racially and ethnically balanced jury could be found. Even then, the judge withheld the decision from the press for more than 4 hours until police could deploy in Miami. Rainy weather, plus 200 National

Guard troops and more than 1,000 police officers, helped maintain calm the next day.

The event precipitating mob violence was, once again, the death of an African American, this time at the hands of a Hispanic police officer. The 1992 Rodney King tragedy resembled the events in Miami in 1980 in that police violence against a black victim was followed by a trial of the alleged perpetrators. Rioting was a deferred reaction to the perceived injustice in each case, due to the acquittal of white police officers. In 1989, however, mob violence was not deferred; it proceeded shortly after the victims' deaths, much in the fashion of 1960s-style rioting. Perhaps this early release of crowd anger explains in part why no further unrest occurred. In any event, the outbreaks that marred Miami during the 1980s carried some disturbing similarities to those arising in Los Angeles in the early 1990s.

The Kerner Commission Remembered

For all their fury, the Miami riots failed to generate similar outbreaks elsewhere in the nation's cities. Little official concern was registered in Washington. Yet there were those who could not forget the lessons of the 1960s. Only 8 months before the Overtown uprising, the House Judiciary Committee had held hearings to mark the 20th anniversary of the release of the Kerner Commission report. The purpose of the gathering was to review the nation's progress in civil rights and social justice in the intervening two decades. Statements were made by Fred Harris, former senator from Oklahoma; Roger Wilkins, an ex-Johnson White House aide; and David Ginsburg, staff director for the Kerner Commission. The United States, they agreed, had made considerable progress in race relations, equal opportunity, and reducing poverty and deprivation since the Great Society era. But they argued that the nation, especially under the conservative policies of the Reagan White House, was slipping farther and farther away from its earlier commitment to justice for minorities and the underprivileged. They echoed the original warning of the Kerner Commission that America was becoming two societies. They bemoaned the "quiet riots" of the 1980s, including family disintegration, housing and school deterioration, segregation, and unemployment.

The hearing report called for a full employment public jobs program, better welfare income maintenance, improved fair housing laws, and advances in school desegregation. More vigorous enforcement of affirmative action and equal opportunity laws was recommended, as well as extended national health insurance and more government support for Head Start and Job Corps programs (*20 Years After the Kerner Commission: The Need for a New Civil Rights Agenda,* 1988). Apparently, few in Washington were listening that day; the warnings uttered at that unusual session quickly faded from public memory.

Two-and-a-half years after the third Miami riot, a less severe incident broke out a few miles from the White House. President George Bush, whose 1989 inauguration had postdated the Miami riot by days, was now in office. The Washington incident occurred on May 5 and 6, 1991. Again, police action and minority response was at the heart of the outbreak. This time, however, the victim was Hispanic and the civil authority was African American. A policewoman shot a criminal suspect in the District of Columbia's Mount Pleasant area, alleging that the man lunged at her with a knife. Bystanders claimed that the Hispanic man had been handcuffed at the time of the shooting. In the rioting that followed, 10 police were injured, 113 people were arrested, and stores along several blocks were damaged. Mayor Sharon Pratt Dixon declared a state of emergency and put the neighborhood under a midnight to 5 a.m. curfew (Ayres, 1991; Krauss, 1991). The nation's capital had not suffered civil disorders since the 1968 riot following the assassination of Martin Luther King, Jr. Although a much less disastrous event, the 1991 outburst proved once again that not even the national capital was exempt from urban interracial mob violence. The melee also demonstrated that violence can erupt from conflict between black police officers and Hispanic residents, just as the 1989 riots in Miami showed that violence can arise from tensions between a Hispanic police officer and black residents. The riots in Miami, Washington, and Los Angeles make clear that conflict will continue to ignite between civil authorities and low- and moderate-income ghetto residents. The simple pattern of conflict between white police officers and young blacks is no longer so uniform; in the future we can expect more violence between and among whites, blacks, and Hispanics.

Los Angeles, 1992,
and the Presidential Candidates

By the time presidential candidates had arrived in Los Angeles in 1992, the cinders had cooled in the burned out South-Central area and city crews and neighborhood volunteers had swept and shoveled most of the streets clear of the debris. Network and local television crews followed candidates Bush and Clinton through their separate walking tours as each candidate shook hands with spectators and federal troops still on guard duty in the city. In no time at all, Bush and Clinton found themselves walking the same tightrope that Lyndon Johnson had walked a quarter century before. On the one hand, the air prickled with the need for official sympathy. Expressions of shock were called for. Los Angelenos, at least those in the riot areas, had been through a lot. Among the president's first official acts was a series of heartfelt speeches and comments to the press showing his dismay. One reporter observed that Bush "seemed stunned by the devastation. . . . 'Things aren't right in too many cities across our country,' " Bush declared to an audience of L.A. citizens. " 'We must not return to the status quo. We must try something new' " (quoted in Beamish, 1992, p. 1).

But like Johnson before him, Bush could not afford to say anything that smacked of sympathy for those who participated in the mayhem. Predictably, he called for tougher law enforcement measures. He also emphasized the need for more individual responsibility among citizens. As the television cameras whirred, the president announced a $19 million grant to the city to enhance its fight against youth gangs and drug dealers. In the face of a billion dollars in damages to the city, this seemed a modest commitment. But Bush was painfully aware of the primary issues in the election campaign—the economy, the need for job growth, and the mounting federal deficit. Instead of using the event to announce costly new federal outlays, Bush took the opportunity to remind citizens of his administration's proposals before Congress for urban enterprise zones, ownership and management of public housing by tenants, and reform of the welfare system. He also called for greater competition among schools through education vouchers that would permit parents to send children to the public or private school of their choice.

When Clinton's turn to face the cameras came, he echoed many of Bush's views. He endorsed Bush administration proposals for urban enterprise zones, selling public housing to tenants, and "weed and seed" projects to stiffen law enforcement while improving social service and job training efforts. Bush viewed these ideas as furthering his philosophy of individual initiative by reducing the involvement of government in local affairs. Clinton, on the other hand, painted them with a different brush. Clinton believed Washington was the appropriate agent for bringing about such changes. This meant that welfare recipients should be offered a public service job after 2 years of assistance. It meant that the feds should capitalize a national network of inner-city community development banks to assist minority small businesses. In addition, Clinton called for more money for early childhood education and assistance to states to equalize school spending between districts (Brownstein, 1992).

In the weeks that followed, a split developed in the Bush administration over a proper response to the riots. For more than 2 years, HUD Secretary Jack Kemp had been calling for congressional passage of an urban aid bill with money to establish an enterprise zones program and the weed and seed proposal. The enterprise zone concept had been a mainstay of the Reagan administration, but had failed congressional enactment three times during the 1980s. Weed and Seed would weed neighborhoods of drug dealers and seed neighborhoods with social services and after-school youth programs. Counting on the riots to give him new leverage with Bush and Congress, Kemp called for enactment of the measures by the end of June. Kemp argued that Bush could open up new support for the Republican Party among the urban poor, minorities, and progressives. Meanwhile, Richard Darman, director of the Office of Management and Budget, was worried about the effect of new urban expenditures on the federal debt. During his tenure, he had quietly urged the president to scale back many of Kemp's proposals. Darman felt that Bush had little to gain politically from expensive new urban policy initiatives. He continued his call for fiscal restraint. Predictably, Vice President Dan Quayle, who attributed the riots to social anarchy and a "poverty of values," pressed the law-and-order theme, a popular refrain among conservatives. In a televised speech to the American public 2 nights after the riots, Bush stressed law and order rather than the concern Kemp and others voiced for new programs to inspire hope in the inner cities (Kranish & Gosselin, 1992). A few days, later Governor

Clinton called for a $24 billion commitment from Congress to implement his proposed urban program; $20 billion of this would be wrested from savings due to cuts in defense spending (Haas, 1992).

With an election breathing down his back, President Bush realized that his administration had to demonstrate some level of support for new urban policies. Perhaps the president took comfort in the results of a *Washington Post* survey, taken the week before, that found that 42% of adults in the nation believed that the social programs begun in the 1960s were a major source of urban problems in the 1990s (Morin, 1992). This, of course, synchronized nicely with a gratuitous comment from Bush's press secretary, Marlin Fitzwater, that the L.A. riots grew out of the failure of the War on Poverty programs of the 1960s. Urban decay, crime, and welfare dependency, Fitzwater insisted, were the result of these failed policies (Morin, 1992). Yet Bush's satisfaction was short-lived when he learned that the survey also found that an even larger percentage of Americans (55%) said that the source of urban problems was the failure of the Reagan and Bush administrations to address urban problems. Even more discouraging was the revelation that 70% of respondents believed that urban problems "are so difficult . . . they will never be solved" (Morin, 1992, p. 37). In the weeks ahead, the Bush administration would fashion a position on the L.A. riots—and more generally, on the federal response to problems of the cities—that appears to have drawn inspiration from these findings. It was clear that official Washington would have to respond with something in the short term.

Note

1. The lawlessness that followed was particularly vicious. Several atrocities, including immolation, repeated stabbings, and battery with chunks of concrete and bricks, were committed by black mobs against innocent whites driving by. Mindful of the horrific cruelties imparted to blacks by white vigilante lynch mobs in an earlier time, the Liberty City rioters ran over victims with cars (Portes & Stepick, 1993). Eleven years later, the Rodney King incident would show similarities to the tragedy befalling Arthur McDuffie; 12 years later, the mob violence in Los Angeles, although apparently not spawning as many bestial acts, would produce the similar tragedy of white truck driver Reginald Denny being dragged from his vehicle and pounded mercilessly by four black youths.

8

Responding to Urban Interracial Mob Violence in the 1990s

If the 1992 Los Angeles riots achieved anything, they lent new visibility and salience to the pleas of the nation's mayors for increased federal urban assistance. Taking advantage of the situation, the U.S. Conference of Mayors, headquartered in downtown Washington, rallied its members for a parade down Constitution Avenue. Although planned months before the L.A. riots, the event was a miracle of fortuitous timing. Amid an estimated 200,000 bystanders, hundreds of city hall officials, homeless people, and urban advocates from around the nation called for $36 billion in new federal funding for cities. Not surprisingly, the danger-of-violence theme was adopted by several speakers. Without significantly greater spending, these speakers warned, the nation would suffer more riots. In terms not heard since the 1960s, New York Governor Mario Cuomo renewed the image of cataclysmic finality. "You can bring this nation together right now to save our cities or stand and watch this nation explode from Los Angeles to New York" (Kranish, 1992, p. 2), he warned. It was vintage 1960s in replay. One observer calculated that the Los Angeles riots had raised the likelihood that the mayors' proposal would be enacted "from the wildly impossible to the merely outlandish" (Haas, 1992, p. 1243).

As George Bush, Bill Clinton, and Ross Perot vied for voter support, the mayors found themselves and their cities receiving newfound attention in Washington and in the mass media. Newspapers and magazines printed extensive coverage of the riots and interpretive articles on background events. Network, cable, and local television broadcasters ran hundreds of news and documentary features. After years of neglect, the cities were enjoying reawakened interest.

Perhaps sensing this, the president adopted a tone of sympathy with community leaders in Los Angeles in a speech he presented there in late May. Although mentioning the premise that all levels of government had to accept responsibility for resolving urban problems, Bush reminded his listeners that Washington had spent billions, maybe trillions, of dollars on cities over the previous quarter century. Although well-intentioned, he argued, most of these programs had simply not worked. Echoing a refrain from his mentor and predecessor, Ronald Reagan, the president insisted that what cities needed most was to get the feds off their backs. Washington, he reminded, encourages dependency, dictates over local prerogatives, and meddles in too many lives.

The primary problem in American cities, Bush said, was the "dissolution and the decline of the American family." To turn cities around, the strength of the family in the lives of individuals must be restored. Although government has a role in this struggle, society itself must help. Here, Bush harkened to a theme from the early days of his administration when he called for a thousand points of light, a euphemism for greater citizen volunteerism in resolving human problems. He concluded by calling for a restoration of society's "moral sense of right and wrong" ("Remarks of President George Bush," 1992, p. E1563).

The Urban Aid Bill in the House, 1992

The first order of business for Congress in the aftermath of the Los Angeles riots was an emergency appropriation to help finance the costs of clearing debris and assisting businesses to reopen. The general attitude in the House was to speed such a measure along to passage in the hopes of responding to the anger and devastation in Los Angeles as quickly as possible. Democrats drafted Craig Washington (D-TX) to deliver an hour-long oration on the House floor embracing the range of views and emotions of members of his party on the riots and appropriate federal responses. He began by denouncing the beatings of Rodney King and Reginald Denny. It was "wrong," he said, for some "to take the law into their own hands and to take out their frustrations, and vent their hostilities on innocent people merely because they happen to have been of another race" ("Aftermath of the Disturbance," 1992, p. H3081). Representative Washington dismissed the need for further studies of the causes of urban violence, noting that the Kerner Commission report of 1968 provided answers still current in 1992. The problems, Representative Washington insisted, were unchanged. He decried the failure of federal policies over the previous decade to provide assistance to cities equal to the job of reducing unemployment, improving housing and schools, and upgrading public health. Representative Washington called for redirecting budget priorities, reducing defense and foreign expenditures, and increasing urban and social outlays. He recommended rethinking the 1990 congressional protocol that agreed to an annual cap on total budgetary commitments, charging that it stifled opportunities to bolster spending on cities and the poor ("Aftermath of the Disturbance," 1992).

Representative Washington called for a congressional response that (a) condemned civil violence based on racial antagonisms, (b) sympathized with the frustrations of citizens who indulged in or were victims of the lawlessness, (c) underscored that the root causes of the tragedy lay in poverty and discrimination, and (d) called for substantial additional federal financing to restore the nation's commitment to resolving these problems.

Whether the L.A. riots should initiate such a nationwide urban campaign continued to be a theme as the House and Senate wrestled with the dilemma. Meanwhile, mindful of the need for immediate aid to Los Angeles, Maxine Waters (D-CA), in whose district the unrest had occurred, and Henry Gonzalez (D-TX), chairman of the House Subcommittee on Housing and Community Development, introduced H.R. 4073, the Gonzalez-Waters Distressed Communities Assistance Act. This act would provide $500 million to guarantee loans by local authorities for riot-related damage in any area "in which a public disturbance involving acts of violence occurred" ("Introduction of Legislation," 1992, p. H3080) between April 29 and May 6, 1992.

Financing could cover new inventory, property acquisition, new construction and rehabilitation, and start-up costs. The bill would back businesses damaged during the rioting and new businesses seeking to locate in the riot areas. The bill was directed primarily toward restoring businesses harmed by looting, arson, and vandalism, but it would also subsidize new entrepreneurs and the expansion of existing firms into riot areas. Those who could not qualify for conventional financing or chose not to do so in the hopes of obtaining lower-cost federally backed loans would be eligible for funds under this proposal.

During the week that followed, at least one representative would find time to insert a bit of levity in the debate over a response to the Los Angeles event. James A. Traficant (D-OH) lamented the loss of manufacturing jobs in America, alleging that 15 million workers were "working for peanuts, below the poverty level" ("American Cities in Danger," 1992, p. H3097). His cynicism barely concealed, he referred to possible federal support for job training:

> But maybe they will be lucky and, with some of the training money, they may be trained as a jelly-roller or as a corncob-pipe assembler or, if they are lucky, they may get a high technology training program as a panty-hose-crotch-closer. (p. H3097)

The Ohio congressman closed on a note of warning about job migration to Mexico and joblessness among young Americans. These conditions, he added, say "a lot why American cities are in danger of all blowing up in flames" ("American Cities in Danger," 1992, p. H3097). Although viewed by many in Congress as hyperbole, Traficant's evocation of the danger-of-violence theme would be repeated by others in less pyrotechnical terminology. But first, Democrats had a grudge to bear. In the day that followed, they would both defend the Johnson War on Poverty legacy and counter Republican criticisms with an appeal to America to revive its support for cities and the poor.

Henry Gonzalez, a fellow Texan and friend of Lyndon Johnson, cited comments by Bush's attorney general, William P. Barr, on April 26 that the problems of inner cities were "essentially the grim harvest of the Great Society" ("Setting a Firm Course," 1992, p. H3226). Barr had argued that the expansion of welfare programs had contributed to the breakdown of family structure; he cited the rate of urban illegitimate births as proof (p. H3226). Gonzalez lambasted Barr, calling his statement "outrageous" (pp. H3225-H3226). He condemned as "offensive, perhaps reprehensible, if not outright despicable" the alleged tendency of presidents Reagan and Bush "to blame somebody else for whatever problems they were not focusing on" (p. H3225).

Gonzalez introduced an earlier statement from Senator Daniel P. Moynihan (D-NY), who had called Barr's words, "an untruth, a lie" ("Setting a Firm Course," 1992, p. H3226). Although serving as Nixon's advisor on urban affairs, former Harvard professor Moynihan had been a young aide in the Johnson administration. In that capacity, Moynihan had contributed to Great Society policy development. The Bush administration, Moynihan had professed, was guilty of "the Orwellian rewriting of history" (p. H3229). Gonzalez asserted that current urban conditions would be even more explosive if Great Society programs had not been enacted. Had those programs been continued, he added, "we would not now have a housing crisis, a return to the high dropout rate, our job-training crisis, and the resultant terrible crime wave we are experiencing" (p. H3229). Jake Pickle (D-TX), another Johnson protégé, echoed his colleague's themes, opining that the events in Los Angeles may not have occurred had Great Society programs been continued or expanded (p. H3231).

Democrats continued to reminisce about the Great Society. It was not long, however, before the danger-of-violence warning reemerged. Reading extensive excerpts from a speech delivered by President Johnson in 1965 at Howard University, Major R. Owens (D-NY) closed with an appeal: "Unless we want to see more urban death and destruction we must reinstitute the war on poverty and stop making war on the poor" ("Setting a Firm Course," 1992, p. H3233). The theme was continued by Howard Wolpe (D-MI), who took the Bush administration to task for "continued insensitivity to deep-seated white prejudices and paternalism" (p. H3234). The nation had a choice, Wolpe warned. It could descend "into the hell of racial conflict and deepening racial polarization," or it could view the misfortune of Los Angeles as "an opportunity to address the underlying causes of the rage and the violence that are erupting all around us" (p. H3233).

Whatever purposes were served by Republican condemnation of the Great Society, the comments brought a torrent of rebuke from Democrats. Gonzalez and Pickle were followed by others sympathetic to their views. Most championed programs such as Head Start and the Job Corps; all called on Congress and the administration to support greater assistance to the urban poor.

Meanwhile, others on Capitol Hill were hoping to capitalize on the momentum in the wake of any Los Angeles emergency response. Behind the scenes, efforts were under way to expand the Gonzalez-Waters bill to include federal support for another tragedy. Preceding the Los Angeles riots by 16 days, flooding in Chicago had brought millions of dollars in damages to downtown businesses. Century-old underground tunnels running under the Chicago River, accidentally punctured by construction workers, brought millions of gallons of water into the basements and streets of Chicago's Loop area. A new bill, H.R. 5132, was cobbled from the Gonzalez-Waters bill proposing funding to repair damages resulting from the Los Angeles riots and from the Chicago flooding. The new measure would not go unchallenged by Republicans.

Although raising no objections to emergency assistance for riot relief, F. James Sensenbrenner, Jr. (R-WI) took H.R. 5132's sponsors to task. He proposed an amendment prohibiting the use of federal funds "to pay for expenses related to cleaning up after the man-made disaster in Chicago" ("Request to Consider," 1992, p. H3174). Sensenbrenner cited "gross negligence on the part of the Chicago municipal government," insisting that this event was "considerably different than the issues presented by providing disaster assistance to Los Angeles" (p. H3174). In opposition, Bob Traxler (D-MI) reminded his colleagues that President Bush had declared both Los Angeles and Chicago disaster areas, thereby automatically qualifying them for disaster assistance. H.R. 5132, he implied, was simply a vehicle for carrying out the president's will.

At issue in Sensenbrenner's view was the fact that taxpayers in other states were being asked to help Chicago finance the costs of its own incompetence and gross negligence. He mentioned media coverage in Chicago that questioned whether the city had failed to maintain its own infrastructure (i.e., the underground freight tunnel that collapsed). Traxler parried with Sensenbrenner over matters of procedure and precedent. Then he reminded his colleagues that in the past week there had been a marvelous sense of comity between Congress and the White House and called on Sensenbrenner to "accommodate the national concerns" ("Request to Consider," 1992, p. H3175). As it turned out, Sensenbrenner's amendment drew little support in a chamber ever aware that the next tragedy could be in one's own district.

Representative Traficant weighed in again, not questioning federal assistance for either city. He worried that "our cities are deplorable, dangerous conditions, ready to explode right in our faces" ("Request to Consider," 1992, p. H3176). Traficant recommended neither raising the taxes nor raising the deficit. Instead, he called for reallocating $7 billion in foreign aid money to America's cities and public schools (p. H3176). Shortly thereafter, C. Christopher Cox (R-CA) rose to drum up support for his Turbo Enterprise Zone bill. Essentially a souped-up version of the enterprise zone measures considered in several previous sessions of Congress, the bill would remove all sales, income, payroll, and property taxes for 5 years on businesses investing in the South-Central area of Los Angeles. (Previous enterprise zone bills had proposed to reduce certain taxes but not to eliminate them.) Cox promised that South-Central would "rival Hong Kong for economic enterprise" (pp. H3176-H3177) once the measure was fully carried out. His damn-the-torpedoes, full-speed-ahead enthusiasm found few converts that day.

The House floor debate on urban issues seesawed back and forth between partisan camps. Romano L. Mazzoli (D-TX) reminded his colleagues that the United States was an urban nation and it needed an urban agenda. He recommended a program of city initiatives proposed recently by the U.S. Conference of Mayors ("Republicans Care About Urban Problems," 1992). The high cost of that package drew little response from Mazzoli's colleagues.

As a California congressman, Dana Rohrabacher (R-CA) found himself in the uncomfortable position of having to bring home the federal bacon to his riot-ravaged state while upholding the Bush administration line. He carped that Democrats had attempted to label the GOP as uncaring about cities and the poor. Democrats, who controlled the House, had stifled Republican policies under President Reagan. Rohrabacher defended the Reagan administration's emphasis on family values and its attempts to develop policies to create jobs, investment in inner cities, and tougher "nonracially biased law enforcement" ("Republicans Care About Urban

Problems," 1992, p. H3177). His admonition drew no retorts from Democrats, who realized that they had the votes needed to clinch the measure.

On the following day, Rohrabacher's California Republican colleague, Robert K. Dornan, added his concern that taxpayers would have "to foot the bill for the willful destruction of property while those responsible go free, their evil actions excused in a fit and frenzy of political correctness" ("Emergency Time-Sensitive Assistance," 1992, p. H3269). He revealed his fears that federal assistance to Los Angeles would convey "a message here that rioting pays and that this Congress can be blackmailed by thugs and vicious, predator gangs" (p. H3269).

Dornan focused on the theme of individual responsibility, rejecting social victimization as an acceptable explanation for the riots:

> The Los Angeles riots were not about jobs or frustration or a feeling of alienation. They were about some people, some individuals—black, white and every hue in between—taking advantage of an emotional situation to steal from their neighbors, knowing full well that liberal politicians and social activists would excuse, even encourage, their actions. ("Emergency Time-Sensitive Assistance," 1992, p. H3269)

Other members of the House soon echoed Dornan's message. David Dreier, yet another California Republican, and James Bunning (R-KY) added their voices to the chorus of doubt about H.R. 5132 ("Emergency Time-Sensitive Assistance," 1992). Both warned that enactment would be tantamount to remunerating those involved in mob violence. The last word on this issue in the House that day came from Porter J. Goss (R-FL). He reminded his colleagues that "we are in no way condoning or rewarding the thuggery, the looting, and the total lawlessness that ensued in Los Angeles" (p. H3273). Instead, the House was simply showing compassion to the victims of the violence. The reward-the-rioters theme would be rejoined a few days later in the Senate, however.

Another critical theme emerged in the House. This concerned the question of federal funding for disaster relief under current legislation when neither event—in Chicago or in Los Angeles—was a "natural" disaster. George W. Gekas (R-PA) first broached the subject. He reassured his colleagues that the Cuban refugee crisis in Florida following the Mariel boatlift and the Love Canal hazardous waste threat in New York state— neither a natural disaster—had received emergency aid from Congress.

Should he ever have to seek similar assistance for a disaster in Pennsylvania, Gekas intoned, he would not want "opposition based on the fact that it was not a natural disaster" ("Emergency Time-Sensitive Assistance," 1992, p. H3266). Thus, he supported H.R.5132. Charles A. Hays (D-IL) agreed, making clear that this was not a partisan issue (p. H3267).

A response immediately followed from Bill Emerson (R-MO), who would support aid for victims of an earthquake, typhoon, hurricane, or drought, for example, but not for "human disasters. . . . Riots. Arson. Looting. Chaos. Killing. All of these were caused by human actors exercising their free will, not [by] natural disasters" ("Emergency Time-Sensitive Assistance," 1992, p. H3267), he reminded his colleagues. Emerson dismissed the anger of rioters over the King jury verdict and insisted that taxpayers should not have to finance "willful misconduct" (pp. H3267-H3268). A rejoinder followed from William O. Lipinski (D-IL), who argued that neither "nature or negligence" (p. H3268) should be at issue as Congress carries out its responsibility to help disaster victims. He championed support for both cities.

Lipinski went one step further, insisting that the distinction was a moot point. He noted that the primary law governing disaster assistance, the Stafford Disaster Relief Act, renders aid regardless of the cause of a fire, flood, or explosion. Thus, the Los Angeles riots and the Chicago flooding, irrespective of whether human commission was a causal factor, were eligible for disaster assistance ("Emergency Time-Sensitive Assistance," 1992).

Later in the day, provisions of the Stafford Act would stir further confusion. Mervyn M. Dymally (D-CA) announced that the law would provide compensation to victims in Los Angeles regardless of whether their property was damaged by fire "or only looted" ("Emergency Time-Sensitive Assistance," 1992, p. H3271). His view was later contradicted by John T. Myers (R-IN), who believed that "looting would not be covered" (p. H3274) by the Stafford Act.

The broader issue of emergency funding for nonnatural disasters arose again that day. Kentucky Republican Jim Bunning ensured House members that the Los Angeles riots were "self-inflicted" and not "natural" ("Emergency Time-Sensitive Assistance," 1992, p. H3273). Tom Lewis (R-FL) agreed that both cities had suffered from a "human mistake" and not from "an unavoidable emergency" (p. H3273). Both men therefore opposed H.R. 5132. Another Republican, Peter J. Goss of Florida, pledged to support the bill but signaled his discomfort. By funding emergency aid for "manmade disasters," he worried, Congress would be "moving beyond the bounds of appropriate disaster relief" (p. H3273).

Like the danger-of-violence and reward-the-rioters themes, the theme of natural versus manmade disaster assistance under the Stafford Act would be revisited on the Senate floor. When a vote was finally called in the House however, H.R. 5132 passed with 244 yeas, 162 nays, and 28 not voting ("Emergency Time-Sensitive Assistance," 1992, p. H3275). The bill provided $494.7 million in supplemental appropriations for small business loans and emergency grants to Los Angeles and Chicago.

The Urban Aid Bill in the Senate

On the opposite side of Capitol Hill, Senate Democrats were revving up support for an amendment to H.R. 5132. It would raise federal spending on several urban programs, extending assistance to all eligible cities in the nation. In addition to being an emergency measure for Los Angeles and Chicago, the House bill would heed the danger-of-violence warnings and become a conduit for increasing generalized urban aid. Blessed by bipartisan cosponsorship from Orrin Hatch (R-UT) and Ted Kennedy (D-MA), the amendment would spend an additional $1.45 billion on Democratic standby programs such as summer jobs for youths, summer Head Start programs for preschool children, and compensatory education for school children. Additional support for Weed and Seed, a Bush administration proposal to weed lawbreakers out of neighborhoods and seed these areas with enrichment services for children, was also included ("Emergency Time-Sensitive Assistance," 1992). Although extending emergency assistance to other cities had been discussed on the House side, the Senate would propel the idea forward.

For Senator Kennedy, the Kennedy-Hatch amendment was a comedown from his proposal on May 6 of a $5 billion measure to increase spending on several urban and welfare programs, including community development, summer youth jobs, Head Start, law enforcement, and emergency food and shelter.

Endorsement came immediately from Christopher J. Dodd (D-CT). Like Congressman Traficant, Dodd resurrected the danger-of-violence argument. Ignoring problems in the cities, he insisted, "will only sow the seeds of future disasters, and engender tragedies such as that of South-Central Los Angeles" ("Federal Assistance," 1992, p. S6977). He cited a litany of urban dilemmas and called for additional spending by Congress.

John Kerry (D-MA) resumed debate on the Kennedy-Hatch amendment.[1] In announcing his support for the Kennedy-Hatch amendment, Kerry emphasized that he was not supporting rewards to those who rioted. This kind of thinking, he noted, comes from the notion that rioting and problems of the urban poor are minority conditions and not of importance to the majority of citizens. Kerry called on the nation and on his congressional colleagues to realize that all Americans—the poor and minorities— had a stake in the future of cities ("Federal Assistance," 1992).

As if to counterbalance Kerry's message, the theme of rewarding the rioters in Los Angeles resurfaced. Mitch McConnell (R-KY) noted that about 40% of felony cases from the rioting handled by the Los Angeles District Attorney's office involved defendants with previous criminal records. Many of the rioters were not motivated merely by discontent over the Rodney King court verdict, McConnell said. He worried that endorsing H.R. 5132 would reward "hooliganism and violence":

> I fear that this bill may be a nearly $2 billion message to the American people: that if you want to destroy your neighborhoods—to loot, rob, and kill—Uncle Sam will reach into his wallet and bail you out. ("Federal Assistance," 1992, p. S6987)

Calling for less welfare assistance, more small business development, and support for enterprise zones and privatizing public housing, McConnell reminded his colleagues that rural areas also needed development assistance. "Above all," he concluded, "we must not reward random violence or incompetent government with federal financial windfalls" ("Federal Assistance," 1992, p. S6987).[2]

Just as it had in the House, the enterprise zone theme arose in the Senate as the emergency aid bill flushed amendment after amendment out of the underbrush. Recognizing that immediate needs in Los Angeles and Chicago would give too little time to argue the merits of a national enterprise zone program, Theodore Lieberman (D-CT) and Bob Kasten (R-WI) asked for a "sense of the Senate" resolution that it pledge itself to adopt zone legislation after the impending congressional recess and before July 4 of that year. Longtime advocates of the concept, the two senators proposed the amendment in deference to the wishes of HUD Secretary and ex-congressman Jack Kemp, a zones zealot since the early 1980s. Not wanting to lose the post-riot steam driving federal urban investment forward, Kemp, widely respected on Capitol Hill, sought to exact a promise from the Senate through Lieberman and Kasten to revisit the issue in the weeks ahead (Sparks, 1992).

No resistance to the Lieberman-Kasten amendment arose, and it was approved. But concern surfaced again that the Senate not appear to be compensating those in Los Angeles who were involved in the violence. John Seymour (R-CA) delivered a sharply worded rebuke to the "hoodlums, hooligans and thugs" ("Federal Assistance," 1992, p. S6997). Citing the number of deaths, injuries, and fire calls, he condemned the lawlessness in Los Angeles. He implored his colleagues not to tie assistance to Los Angeles to a national urban aid bill. Seymour cited several natural events designated disasters in California, lumping the riots with them. Although castigating the looters and arsonists, the California Republican did not want the Senate to saddle the fundamental issue—emergency disaster assistance—with the burden of a floor debate on national urban policy. Whether urban rioting should be categorized with natural events went undiscussed.

Later in the floor debate, Seymour presented another amendment to H.R. 5132 (1848). This amendment would forbid aid expenditures on those arrested, convicted, or "subject to pending charges" ("Federal Assistance," 1992, p. S7006) for riot-related crimes in Los Angeles. Extolling the virtue of those who did not engage in unlawful behavior, Seymour cautioned his colleagues not to "reward rioters, looters, arsonists, and murderers" (p. S7006). Phil Gramm rose to speak in behalf of Seymour's measure. Describing a meeting with Korean Americans from the Los Angeles riot area, the conservative Texan tried to distinguish funding for relief after natural disasters, such as floods and earthquakes, from funding for relief due to human-caused events such as assault and battery, arson, looting, and vandalism during the Los Angeles riots. In natural events, there is effectively no human culpability, Gramm argued. There is no chance then that

federal assistance will be misspent on those who contributed willfully to
the calamity. In rioting such as that in Los Angeles, however, the risk is
too great that the guilty will benefit unjustly.

Calling the rioters "hardened criminals" ("Federal Assistance," 1992,
p. S7013), Gramm endorsed Seymour's amendment, insisting that "none
of the taxpayers' hard-earned money is going to go to people who took
advantage of the situation, who burned and looted and killed in Los
Angeles, California" (p. S7013). As it turned out, the Seymour amendment
carried, 68 to 28. A substantial majority of the Senate wanted to make it
clear to the press and the American public that Congress would not reward
the rioters.

Discussion on the "manmade" nature of the Los Angeles riots spurred
Illinois Democrat Alan J. Dixon to speak out on the Chicago disaster
assistance in H.R. 5132. He acknowledged that neither Los Angeles nor
Chicago has suffered "natural occurrences, such as tornadoes, hurricanes
or earthquakes" ("Federal Assistance," 1992, p. S7016). Yet he sought
sympathy for the many businesses and city facilities in Chicago's Loop
suffering damages exceeding $50 million. Then, as if to confirm Sensen-
brenner's earlier charges of municipal negligence on the House side, Dixon
noted that the collapsing tunnel system was old and in poor repair. Directly
or indirectly, it seemed, H.R. 5132 would assist Chicago in repairing the
damage. There was no mention of municipal or state liability when a
publicly owned capital facility fails, causing massive damage to private
property. Nor was there mention of the fiduciary responsibilities of private
insurers to cover those damages.

The distinctly nonnatural character of the disasters in the two cities in
question persisted. Barbara Mikulski (D-MD), who chaired the Appropriations Subcommittee with oversight on disaster assistance programs, explained an obscure but critical point to those on the Senate floor. She
reassured California Senator Seymour that H.R. 5132 would provide
assistance to those affected by riot acts such as looting and vandalism and
the fires. The law makes "no distinction in the administration of Federal
disaster relief programs between fire and non-fire victims of the Los
Angeles disaster" ("Federal Assistance," 1992, p. S7017).

Seymour was concerned that, under federal disaster assistance law,
only damage from a "natural" phenomenon such as fire would be covered
by H.R. 5132. Damage due to rock throwing, pillaging, firearms dis-

charges, and the like—undeniably human acts—would not be included, he feared. Mikulski indicated, however, that Federal Emergency Management Agency officials had interpreted their authorizing legislation "broadly to be inclusive rather than exclusive" ("Federal Assistance," 1992, p. S7017). In other words, urban mob violence would be considered in the same league as natural disasters, as far as federal emergency assistance policy was concerned.

A week after the House vote, the Senate passed H.R. 5132, now grown from a $.5 billion to a $1.5 billion measure. The Kennedy-Hatch amendment, extending aid beyond Chicago and Los Angeles to many cities, accounted for most of the increase. In conference committee, the bill was trimmed to $1.1 billion, largely due to worries of the effect of the measure on the federal deficit and to feelings that it awarded too much to cities and not enough to rural and suburban communities. On June 18, both houses approved the measure ("Urban Aid," 1992). A day later, President Bush signed the legislation (Eaton, 1992).

As enacted, the Supplemental Appropriations or Urban Aid bill contained $256 million for the Small Business Administration to be used for small business and disaster relief loans and grants; $500 million for Head Start, Weed and Seed, and summer school programs; and $300 million for disaster clean-up and assistance work by the Federal Emergency Management Agency in Los Angeles and Chicago. A little more than one half of the money was earmarked for emergency relief. The remaining amount— roughly a half billion dollars—outflanked Congress's budget cap, powered by the rationalization that to fail to subsidize America's cities was to invite further interracial mob violence.[3]

When the emergency appropriations bill was enacted in mid-June, many on Capitol Hill breathed a sigh of relief. In a few weeks, the House would approve a broader urban aid package, including the enterprise zone initiative. Meanwhile, Tom Campbell (R-CA) quietly introduced the Riot Reinsurance Act of 1992. A federal riot reinsurance program had existed from 1968 to 1984; Campbell's measure would have resurrected the system. To encourage businesses to locate or remain in inner cities, Campbell explained, investors need indemnification should losses occur due to civil disturbances such as those in Los Angeles. Under the measure, private insurers of urban businesses would receive additional financial backing from Washington to help cover damages in the event of a civil disorder. Washington thus would share in the risks of inner-city businesses due to mob violence. In turn, the federal program would share in the premiums paid by the insured.[4]

The Enterprise Zones
Legislation in the House

After the House returned from its recess, members were obligated to honor their earlier commitment to revisit the enterprise zone measure by the recess scheduled to begin on July 4. There was little time to debate in depth the broader dimensions of federal urban policy. But enactment of an enterprise zone bill was possible because in one form or another, the zones concept had bounced around Capitol Hill for several years. Supported by the Reagan and Bush White Houses, enterprise zones were familiar to many members of Congress by 1992. Moreover, well over one half of the states had some version of a state enterprise zone program in operation, offering relief under state and local tax codes. Considerable support existed on both sides of the partisan aisle. In short, enterprise zones were on the shelf, readily available, and would require little in the way of policy education for members of Congress, lobbyists, or constituent groups.

On July 2, the House approved H.R. 11 356 to 55, with 198 Democrats and 158 Republicans voting affirmative. The measure provided for 50 enterprise zones over 5 years funded at $5 billion ("Unanimous Consent Agreement," 1992). One half of the zones would be designated in urban areas and the other half in rural communities. Eighty percent of the benefits would be assigned to the urban zones. The measure prohibited corporations from participating so that small businesses, which generate most new jobs in the nation, could benefit. Only neighborhoods with unemployment 1.5 times the national average would be eligible for zones. In addition, at least 20% of zone residents would have to have household incomes below the poverty line. Eligible investors in enterprise zones would reap savings on several tax obligations:

- Exemption of 50% of capital gains from sale of an enterprise zone business property held at least 5 years; this would cut from 28% to 14% the maximum tax rate paid by these investors

- A credit of 15% on wages paid by employers of residents living in the zone, up to a maximum of $3,000 per employee annually
- Deductions up to $20,000 during the first year of business operation on newly acquired equipment
- Purchasers of stock in zone companies could receive an annual deduction of up to $25,000, with a maximum total deductions per taxpayer of $250,000

The enterprise zone sections were attached to a complicated $14.5 billion tax bill. The $5 billion for the zones would be split, 50% in direct subsidies (doled out over 5 years) and 50% in estimated taxes forgone (over 6 years). The $500 million in subsidies would be used to fund weed and seed programs to support the zones. These would help localities weed out crime by funding aggressive police enforcement against drug dealers and gangs. Efforts to seed positive influences such as Head Start and drug rehabilitation services would follow (Krauss, 1992).

The *New York Times* reacted to the measure with grudging support, editorializing that too little money was being spread over too many cities ("The Muddled Model Cities Model," 1992). The editors implied that Lyndon Johnson's initiative should have taught Congress a lesson about diluting the effectiveness of new programs by simply tossing every state a bone.

The Enterprise Zones Bill in the Senate

Shortly after the House passed H.R. 11, another version of an urban assistance package began to emerge from the Senate Finance Committee, chaired by Texas Democrat Lloyd Bentsen. An amendment to H.R. 11, the new measure contained provisions for a total of 25 zones in both urban and rural settings and would offer $2.5 billion in federal tax incentives. Fifteen zones would be located in urban areas, eight in rural areas, and two on Indian reservations. Distressed at the smaller numbers of zones (as compared to H.R. 11's 50 zones), Joseph Lieberman (D-CT) and Bob Kasten (R-WI) offered a friendly alternative bill. The alternative would provide for up to 300 zones at about the same cost, $2.5 billion, as the Senate Finance Committee bill. This would be accomplished by reducing zone size: Each zone would be populated by less than one third the number of people allowed per zone in H.R. 11 (12,000 versus 43,000). The Lieberman-Kasten measure remained a gleam in its sponsors' eyes, but it helped send the message that 25 zones was insufficient.[5]

By August 11, the Senate had quintupled the original 25 zones provided for in the Senate Finance Committee bill. Seventy-five would be urban, 40 rural, and 10 on Indian reservations ("Enterprise Zones," 1992, p. S12249). To accommodate the higher number of participating communities and Indian reservations, the bill's price tag had risen from $2.5 billion to $5.5 billion (p. S12249).

Bentsen reported to the Senate that the revised measure also would reduce tax credits on wages paid by zone employers, add tax credits for employers who train zone residents for jobs, and reduce the partial exemption of taxes on building and equipment investments by small businesses. Deductions would be reduced for those who purchase stock in zone businesses, and fewer businesses would be eligible to offer the stock deductions. Finally, the revised measure disqualified farms with assets over $500,000 from participating in the rural zones, limiting incentives to smaller family farms ("Enterprise Zones," 1992, p. S12247).

A fundamental dividing line between most Democrats and Republicans was the issue of the availability and size of capital gains relief for investments in zone buildings and equipment. From President Bush to Republican Leader Robert Dole (R-KS) to countless others in the G.O.P., capital gains liberalization was the Holy Grail of tax reform. For many Democrats, however, it was perceived as a giveaway to the rich with few benefits to the poor or middle classes. Consequently, some Democrats fashioned an amendment to the Senate bill that was viewed by Republicans as overly generous. Supported by Ted Kennedy, Donald Riegle, Jim Sasser, George Mitchell (D-ME), Joseph Biden (D-DE), and Harrison Wofford (D-PA), the measure stood on the principle that "tax breaks alone are not enough" ("Tax Enterprise Zones Act," 1992, p. S15026). It would match tax relief dollar for dollar with direct federal spending for programs directed to the needs of those living in the zones. Kennedy described the proposal as a "head-to-head competition between tax expenditures and direct expenditures" (p. S15026). To no one's surprise, the amendment was consistent in philosophy with the earlier measure sponsored by Kennedy and Hatch to include direct expenditure assistance in the Urban Aid bill.

The Democrats' amendment would provide $300 million a year in grants within the zones to support job development and training, housing rehabilitation, education, health care, and law enforcement. Recipient communities would choose from a list of five "menus" of programs to address the problem areas where the greatest local needs exist. The $300 million would increase funding for a host of existing federal programs such as Head Start and Job Corps. Kennedy noted that the amendment would also require a careful evaluation of outcomes in both tax relief and direct subsidy programs to help guide future policy development.

In addition, the measure would award $200 million in grants to cities without federal enterprise zones to fund public-private partnerships between the federal government and community-based organizations. Again, funding would be channeled through existing programs such as Head Start and Job Corps, but $65 million would be earmarked for new programs to offer grants to nonprofit organizations and community development corporations to subsidize small business development ("Tax Enterprise Zones Act," 1995, p. S15031).

Senator Riegle reminded his colleagues once again of the danger-of-violence rationale for greater urban and poverty expenditures. Pointing to the Los Angeles riots, he warned, "I am afraid that, if we do not act quickly with more resolve and more forcefulness, we will see more Los Angeles in communities across the Nation" (p. S15042).[6]

While debate continued over the exact makeup of an enterprise zones package, the issues were dwarfed in fiscal magnitude by the larger cause to which any Senate urban assistance would be grafted: tax relief for middle- and upper-income households and businesses. Hoping to enshroud assistance to cities in a bill with broader appeal, the Senate had joined its amended version of the House-approved H.R. 11 to the Tax Revenue Act of 1992. The total price tag for this omnibus measure had swelled to $31 billion.

Included were $2,500 tax credits for certain home buyers; repeal of luxury taxes on yachts, jewels, and furs; and tax incentives to spur business expansion around the nation. The plum, however, was a clause permitting purchasers of Individual Retirement Accounts (IRAs) to deduct the cost from their income taxes. Echoing a measure eliminated in the 1986 Tax Reform Act, the new IRA provision was meant to stimulate savings and increase capital for investment.

LIVERPOOL JOHN MOORES UNIVERSITY
LEARNING SERVICES

However well-intentioned these tax breaks were, they were largely extraneous to the issues raised by the Los Angeles riots. *USA Today* called the bill "outrageous" and a "giveaway to the well-connected" ("Insiders Win, Cities Lose," 1992, p. 6A). It noted that only one tenth of the tax relief provided by the bill would benefit the enterprise zones. Meanwhile, the IRA provision by itself would add $4 billion to the federal deficit, the newspaper chided. It labeled the bill "a giveaway to the well-connected" and advised the Senate to "kill this turkey" (p. 6A) and get on with the job of devising meaningful urban programs. The *Washington Post* looked askance at several real estate tax breaks, including passive loss deductions and tax credits for first-time home buyers (Mufson & Crenshaw, 1992). Other measures provided relief for retirees, parents paying tuition for college students, and restaurant owners.

Cynicism about the Tax Revenue Act was not limited to the press or to the American public. On August 12, Representative Owens unleashed an acid tongue on the House floor, calling the measure "an orgy" ("The Great Tax Conspiracy," 1992, p. H8132) of spending. With a projected cost of $32 billion, Owens labeled it "obscene" and "a conspiracy" (p. H8132). His ire was targeted at the enormous expense of tax subsidies to business and middle-class beneficiaries, in the face of only $2.5 billion in aid for cities and the poor. Proud that he had voted against H.R. 11, Owens said, "I can think of no plot more dastardly" (p. H8134) than using the Los Angeles riots as an excuse to put forth tax relief for the affluent disguised as an urban aid bill.

As the Senate plodded toward finalizing its bill, the president's task force weighed in with the results of its study of the Los Angeles riots in early August. Cochaired by Deputy Secretary of Education David T. Kearns and Deputy Secretary of Housing and Urban Development Alfred A. DelliBovi, the group recommended creation of neighborhood opportunity centers to provide information and assistance on federal programs to citizens and organizations. It also proposed deferral of loan repayments to the Small Business Administration for up to 15 months for businesses that reinvested in the riot area. In addition, the report called for military-style career academies at public schools to provide vocational and technical training to youths at risk of joining gangs. Other proposals included simplifying federal, state, and local regulations that unnecessarily complicated the rebuilding effort and better coordination of relief programs and services. The report quickly met with criticism from city officials and community leaders in Los Angeles, who argued that it was too vague, did not provide needed funding, and merely repackaged existing programs (Rivera, 1992).

Unlike the impact on urban legislation of the Wood task force 26 years previously, the Bush White House task force had little effect on the Tax Revenue Act. In late September, the Senate approved the measure, 70 to 29, (*Congressional Record,* 1992b, p. S15604). Next, it went to a conference committee, where House and Senate members struggled to reconcile differences between the original House-approved H.R. 11 and the newly passed Senate Tax Revenue Act.

The conference committee revisions were hurriedly completed and the measure went back to the House for a final vote. In the early morning hours of October 6, with the presidential campaign and House and Senate elections in their final month, some House members were understandably tired and irritable. Several complained. Gerald B. H. Solomon (R-NY) charged that "not one man or one woman in the Chamber . . . knows what is in this bill" ("Conference Report," 1992, p. H11633). Some estimated that the omnibus bill contained more than 2,000 pages. Phillip M. Crane (R-IL) called the session "a charade" and "total nonsense" (p. H11636). John R. Kasich (R-OH) termed it "absurd" (p. H11635). Others warned that there were so many tax increases or tax reductions that the bill would come back to haunt Congress when the full effects were apparent to the press and taxpayers (p. H11637). It was not the "greatest thing since sliced bread," said Dan Burton (R-IN), but rather it was "baloney" (p. H11637). Those supporting the measure countered that it was revenue centered, with tax hikes offsetting tax breaks.

As in debates over War on Poverty legislation a quarter century before, medical metaphors arose. For Charles B. Rangel (D-NY), the Tax Revenue Act was not an urban aid bill. It was "not even a Band-Aid" ("Conference Report," 1992, p. H11638) to assist poor neighborhoods, he felt. Yet he would support it because it was at least something. Continuing his doctorly critique, Rangel warned of the "cancer" of urban poverty. He unearthed the danger-of-violence rationale, insisting that the bill, with all its liabilities, "gives us the opportunity not to wait until our inner cities blow up before we come in with billions of dollars and the National Guard to see how many people that we can put in jails" (p. H11638).

Others employed alternative images. Robert H. Michel (R-IL) said that the urban aid portion of the legislation was "lost amidst innumerable, unidentified flying tax objects" ("Conference Report," 1992, p. H11648). He called it a "maze inside a labyrinth surrounded by a jungle—impenetrable to normal human beings" (p. H11648). Noting that only about 10% of the measure's cost would be devoted to urban aid, Vic Fazio (D-CA) called it "a bonanza to wealthy investors" (p. H11655).

It was after 3:30 a.m. when a floor vote was called; it was apparent to everyone that the tally would be close. The measure slipped by, 208 to 202, with 23 not voting. Two days later, the Senate easily passed the measure, 67 to 22. Held responsible by House members for having bloated the bill with billions in tax breaks, the Senate needed little time to endorse it. The new Urban Aid Program would provide for 50 enterprise zones, half in cities and half in rural areas. Although the tax relief measures provided for in the original House bill remained largely unchanged in the final version, the total cost of the measure had skyrocketed from $17 billion to $36.6 billion ("Congress Adjourns," 1992). Throughout the month of October, the bill would be parked on the president's desk awaiting his signature.

As the fate of the Tax Revenue Act played out over the summer and early fall of 1992, a much more compelling drama occupied center stage in the national press. In June, Arkansas Governor Bill Clinton had captured the Democratic nomination for the presidency at the party's convention in New York City. George Bush easily won his party's renomination at the GOP convention in Houston, and the two candidates were soon engaged in political jousting. The central issue in the election was the worst economic recession since the 1950s. Reflecting on rising unemployment rates and business failures during the early 1990s, both candidates placed considerable attention on methods to stimulate the economy. A centerpiece of the Clinton strategy was a call for a national economic plan and programs to "grow" new enterprises and create jobs. Neither candidate proposed Keynesian-style massive federal spending à la New Deal era. But deficit reduction, reallocation of Pentagon funds to domestic needs, and tax concessions were frequent topics of debate.

Bush continued his Reaganian chant for more deregulation and further cuts in federal spending. The Clinton forces rejoined, calling for national health care reform and more federal support for job training, education, and "welfare-to-workfare" programs. Although occasional references were made by each candidate to the Los Angeles riots and to the problems of cities, it was clear to both that there was little mileage to gain by dwelling on these matters. The American public by and large was not in a generous mood; with the nation embroiled in what some were calling its worst economic crisis since the Great Depression, there was little tolerance for new federal investments in poor neighborhoods.

To the surprise of few, the president made no move to sign the Tax Revenue Act prior to the general election. With its huge list of tax relief gifts and widespread mass media criticisms, the modest commitment to enterprise zones was lost in the sauce of politics. Bush realized that endorsing the bill would give Democrats ammunition to charge that the White House was interested largely in the well-off, who would benefit most from tax relief measures. Moreover, there were few Republican votes to harvest in urban or rural poverty areas that might benefit from an enterprise zones program.

When the 1992 election was over, Clinton, with 43% of the vote, defeated Bush (38%) and Ross Perot (19%), ending 12 years of White House domination by Republicans. Bill Clinton would become the second Democrat in 24 years to occupy 1600 Pennsylvania Avenue. As Clinton campaigners celebrated, George Bush quietly vetoed the Tax Revenue Act. Enterprise zones, as well as any other significant urban initiatives, were dead for another year.

• Empowerment Zones and Enterprise Communities Bill

In the months following the April 1992 Los Angeles riots, the enterprise zone measure was propelled forward in Washington on the strength of at least three concerns: the anxiety of members of Congress from urban districts and states over the possibility of more violence, the continued attention of the mass media on post-riot events in Los Angeles, and the pressures of urban, minority, and poverty interest groups in Washington. By the same token, other forces conspired to counter this movement: the national economic recession, high unemployment, and a monumental federal deficit. On top of these, there was a residual public attitude that innumerable federal programs in the past had failed to revive the older cities and uplift the poor; what, then, was the value of one more urban program?

Immediately after the Los Angeles riots, countless articles about the plight of the urban poor appeared in newspapers and magazines. Several television and radio stories and documentaries were broadcast. By the time of Bill Clinton's inauguration in January 1993, however, the burst of attention to the urban crisis had waned. Unemployment continued to occupy center stage in America. Clinton knew that his first order of business would be to reestablish confidence in the economy and to reinvigorate business growth. In the early months of his administration, he pressed themes such as job development, the federal budget deficit, and health care reform. The appointment of Henry Cisneros, former mayor of San Antonio, as the new HUD secretary was an indication, however, that Clinton thought highly enough of his urban constituents to appoint someone of national stature who had confronted their issues first hand. Moreover, as a Hispanic, Cisneros brought further support from ethnic and racial minorities. Appointment of liberal academics such as Donna Shalala as secretary of health and human services and Robert Reich as secretary of labor were additional indications of the former Arkansas governor's sympathies toward cities and the poor. Four other cabinet secretaries—Mike Espy (Agriculture), Hazel O'Leary (Energy), Ron Brown (Commerce), and Jesse Brown (Veterans Affairs)—were African Americans. They, too, symbolized Clinton's ties to traditional Democratic voting blocs.

The debates in Congress over both the emergency urban aid and Tax Revenue Act legislation proved not to be hollow events. Liberal Democrats such as Donald Riegle, Ted Kennedy, William Bradley, John Kerry, Charles Rangel, and Maxine Waters had many urban constituents who would not be forgotten. Moreover, several Republicans who had embraced the enterprise zone concept—backed by two Republican presidents—could not now turn around and oppose it merely because a Democrat was occupying the White House. But for the ill-fated strategy of wedding urban assistance to a fat tax relief package, the enterprise zone bill might have survived President Bush's pen. Many still on Capitol Hill had committed themselves to enterprise zones or other aid to the cities during the 1992 session of Congress. Some wondered if the new administration would take advantage of the opportunity before the Los Angeles riots became a dim memory. It was probably no accident that the White House introduced its own enterprise zone bill just after the first anniversary of the Los Angeles riots. The timing was a response to a host of press articles heralding the date and noting that little had been accomplished over the past year to address conditions that contributed to the violence. The new bill was also an affirmation of Clinton's support for enterprise zones (Lemann, 1994).

Press attention to the urban issue was not the sole source of the Clinton administration's move. Clinton had received a visit in March from former president Jimmy Carter, who urged him to assist private organizations such as community development corporations to fight housing, jobs, and health problems ("Carter Tells of His Fears," 1993). In April, HUD Secretary Cisneros had met with Senator William Bradley, who had emerged as one of Congress's most articulate spokespeople on cities and poverty (Broder, 1993). The two compared notes on strategies and Bradley briefed the secretary on measures that he and a coalition of Republican and Democratic members of the Senate had announced a few weeks earlier. Echoing the menu selection approach of an earlier Ted Kennedy bill, the Bradley Urban Community-Building Initiative incorporated several bills that would allow distressed cities to select program combinations individually tailored to each locality's circumstances. Clinton officials, responding in part to the Bradley initiative, soon realized that the pressure was on to demonstrate their allegiance to an urban agenda. The result was the Economic Empowerment Act of 1993.[7]

The act proposed designation of 100 *enterprise communities* and 10 *empowerment zones*. Sixty-five enterprise communities would be in urban areas and 35 would be in rural regions. Six zones would be in cities, three in rural areas, and one on an Indian reservation. A larger proportional share of resources would go to the zones than to the communities. (I refer to enterprise zones and empowerment communities collectively as EZECs.)

Changes in terminology notwithstanding, the Clinton measure was quite similar to those proposed a year earlier on Capitol Hill. It thus capitalized on existing political agreements and understandings. Over 5 years, almost $4.1 billion in tax reductions would be available to employers locating or expanding in an EZEC. Most of these subsidies would be available as tax credits for training and employing EZEC residents. A small proportion of the $4.1 billion would be reserved to finance direct grants to EZECs for community policing and supportive social services. The budget for the new proposal was twice the size of the enterprise zone measure passed by Congress in 1992. Moreover, it provided for more than twice as many community recipients. Incorporated into the Omnibus Budget Reconciliation Act of 1993 (H.R. 2264), the measure was passed in the House by a vote of 219 to 213 on May 27.

By June 1993, the EZEC bill had drawn fire from several quarters. Some in the Senate felt that too much funding would go to only 10 empowerment zones and not enough to the 100 enterprise communities. The perennial Republican call for capital gains tax relief in the zones, absent from the Clinton bill, was also a point of contention. Resistance from Indian tribes arose because so few reservations would be eligible for a zone or community designation ("Sense of the Senate Resolution," 1993).

Enterprise zones disciple Joseph Lieberman (D-CT) pointed out that the measure spread too little funding over too many designated enterprise communities and too much over too few empowerment zones. Many small entrepreneurs in the more richly funded empowerment zones simply could not use the $320 million per zone promised by the measure, Lieberman argued. Better to reduce the per zone subsidy and distribute it to more empowerment communities. In this way, more cities such as Hartford and Bridgeport could be eligible to compete for a designation. Like many of his colleagues on the Republican side, the Connecticut Democrat called for capital gains incentives to match the wage tax credits in the administration's Economic Empowerment bill ("Sense of the Senate Resolution," 1993).

Given the momentum in Congress to reduce the deficit, incorporating the empowerment legislation into the Omnibus Budget Reconciliation Act at first appeared to be an adroit strategy. As the summer progressed, however, it became apparent that the Budget Act had become a watershed for the new administration. Republicans in both congressional chambers lined up in solid opposition. Several Democrats, skittish about spending cuts in defense, social security, Medicaid, and Medicare and about tax increases on the wealthy were unenthusiastic or downright opposed to the measure as written. Although nearly all members of Congress were publicly for deficit reduction, privately most sought to stave off a budget cut or tax increase affecting key constituent groups.

Buried in the stultifying language of debt reduction, however, were the sections of the bill on EZECs. On action by the Senate Finance Committee, chaired by Senator Moynihan, the EZECs section of the Budget

Act was deleted (Lemann, 1994). By a vote of 50 to 49, the Senate passed a version of the budget measure on June 25, 1993. Only a tie-breaking vote from Vice President Al Gore had saved the legislation. When the two bills were taken up by a conference committee, Republicans in both houses remained solidly opposed to them. To counter their weight, the White House lobbied furiously among the few Democrats who were opposed or uncertain about their vote. For the most part, opposition to the conference committee bill centered on various budget cuts or tax increases. The EZECs sections, strongly supported by the Congressional Black Caucus, were quietly grafted back onto the conference committee bill. It was at this point that Congressman Charles Rangel and Senator Bill Bradley succeeded in expanding the social services block grant portion of the measure to $1 billion (Lemann, 1994).

In the end, the conference committee bill was passed by the House 218 to 216 (August 5) and by the Senate 51 to 50 (August 6), with Gore again making the tie-breaking vote. On August 10, President Clinton signed H.R. 2264 (Pub. L. 103-66) into law at a White House ceremony. *Congressional Quarterly* noted that, for the first time since the end of World War II, the majority party was able to pass major legislation in the complete absence of support from minority members (Hager & Cloud, 1993). Although the EZECs section of the Omnibus Budget Reconciliation Act was never the most critical issue for most of those voting on the legislation, its enactment came by the thinnest of margins. After 11 years of gestation, a federal enterprise zone program had finally been delivered.[8]

Federal Response to Urban Poverty and Mob Violence, 1990s Style

Subchapter C, the EZEC section of the Omnibus Budget Reconciliation Act of 1993, provides a total of $4 billion in tax incentives and $4 billion in grants. The measure sets forth a fairly straightforward program concept, but entangles it somewhat in a web of eligibility qualifications. It provides for a maximum of 65 urban and 30 rural enterprise communities and a maximum of 6 urban and 3 rural empowerment zones. All EZECs were to be designated by the Department of Housing and Urban Development or the Department of Agriculture between 1994 and 1996; each zone or community would receive federal assistance for a maximum of 10 years. Applicants for an EZEC designation were to be either an appropriate state or local government or a state-chartered economic development corporation.

The legislation stipulates complicated eligibility criteria, including population, economic distress, and geographic size. For example, local government applicants for an urban empowerment zone or enterprise community had to have a population of between 50,000 and 200,000, although exceptions exist. Rural local government applicants could have no more than 30,000 people. Urban localities could not exceed 20 square miles in area and rural localities could not exceed 1,000 square miles. Urban empowerment zones and enterprise communities could be located in two states; rural units could be located in up to three states. Central business districts (CBD) could not be included unless the poverty rate in the CBD was 35% in an empowerment zone and 30% in an enterprise community.

The new program required that each EZEC contain "pervasive poverty, unemployment and general distress" (Section 1392 (a) (2)) as a condition of selection for participation. Restrictions mandated that a minimum of 20% of the households in each census tract in an empowerment zone or enterprise community have incomes below the poverty level. From 50% to 90% of the tracts had to have even higher rates of poverty (up to 35% under the poverty line). The HUD and agriculture secretaries were given discretion to lower poverty thresholds for enterprise communities.

Keeping in mind that the total of 104 EZEC awards would ensure that many of the 435 House members would not be able to deliver a designation to their congressional district, it is clear that the elaborate eligibility criteria specified in the EZEC legislation intended to favor certain districts. Key supporters of the legislation were among the favored. Moreover, by giving the HUD and agriculture secretaries discretion to devise additional selection criteria, Congress ensured that a vehicle for bringing political pressure to bear—thereby influencing the outcome—would be available.

Eligible employers in empowerment zones can receive up to a 20% tax credit on wages up to $15,000 paid to each qualified zone employee from 1994 through 2001. After that, the size of the credit will decline until 2005, when it will no longer be available. Employees are required to live in and work in the zone. To qualify, at least 35% of an employer's workers must be residents of the EZEC. Financing for business facilities in the

EZEC is available from tax-exempt facility bonds, which can be issued in an amount up to $3 million per EZEC. The appropriate cabinet secretaries select designees on the basis of the effectiveness of a required strategic plan that each applicant has to submit. The document must include a development plan for the zone, provide for public and institutional participation in developing and implementing the strategic plan, and describe nonfederal resources to be brought to bear. In addition, the strategic plan must identify funding sought from any other federal programs and provide a method for evaluating the success of the program once implemented. (Economic self-sufficiency is a key evaluation criterion.) The plan is expected to demonstrate that EZEC employers will not simply reduce employee rolls at branches located outside the EZEC to compensate for hiring in the EZEC, nor close such a branch and relocate it to the EZEC. In choosing among applicants for an EZEC designation, each cabinet secretary is free under the EZEC program to specify additional criteria.

Sixteen months passed from enactment of the EZEC program to the announcement by the White House of communities successfully competing for designation as an empowerment zone or enterprise community. On December 21, 1994, President Clinton released the names of 104 urban and rural communities, each of which would participate in Washington's newest response to poverty and mob violence. Out of a total of 520 applicants, one in five succeeded in landing an EZEC award. As specified in the legislation for urban EZECs, six empowerment zones and 65 enterprise communities were named (Table 8.1). The Clinton administration added two *urban supplemental zones* to the list of winners by cobbling together funds from other HUD programs. Out of a total of 293 urban applicants, 73 (one in four) received some form of designation.

Of the eight urban empowerment zone designees, all but Los Angeles are located in an eastern or Midwestern state. Of the 65 urban enterprise community designees, 30 are in eastern or Midwestern states. Forty-two of the 50 states, plus the District of Columbia, received at least one EZEC award (communities in Kansas and Missouri shared an award). Alaska, Hawaii, Idaho, Maine, Montana, North and South Dakota, and Wyoming have no recipient communities, possibly due to the failure to apply for a designation. The three largest states—California, New York, and Texas—each have five local government recipients. Michigan, Pennsylvania, Ohio, and Massachusetts—all Rustbelt industrial states—received three apiece.

TABLE 8.1 Communities Receiving an Empowerment Zone or Enterprise
Community Designation, 1994

Empowerment Zones	Supplemental Empowerment Zones	Enhanced Enterprise Communities
Atlanta, GA	Los Angeles, CA	Boston, MA
Baltimore, MD	Cleveland, OH	Houston, TX
Chicago, IL		Kansas City, MO/KS
Detroit, MI		Oakland, CA
New York, NY		
Philadelphia, PA		

Enterprise Communities

Akron, OH	Minneapolis, MN
Albany, GA	Muskegon, MI
Albany, NY	Nashville, TN
Albuquerque, NM	New Haven, CT
Birmingham, AL	New Orleans, LA
Bridgeport, CT	Newark, NJ
Buffalo, NY	Newburgh-Kingston, NY
Burlington, VT	Norfolk, VA
Charleston, SC	Ogden, UT
Charlotte, NC	Oklahoma City, OK
Columbus, OH	Omaha, NE
Dallas, TX	Ouachita Parish, LA
Denver, CO	Pittsburgh, PA
Des Moines, IA	Portland, OR
E. St. Louis, IL	Providence, RI
El Paso, TX	Pulaski Co., AR
Flint, MI	Rochester, NY
Harrisburg, PA	San Antonio, TX
Huntington, WV	San Diego, CA
Huntington Park, CA	San Francisco, CA
Indianapolis, IN	Seattle, WA
Jackson, MS	Springfield, MA
Las Vegas, NV	St. Paul, MN
Louisville, KY	St. Louis, MO
Lowell, MA	Tacoma, WA
Manchester, NH	Tampa, FL
Miami, FL	Waco, TX
Milwaukee, WI	Washington, DC

SOURCE: *News Release: Statement of Henry Cisneros on Announcement of Empowerment Zones and Enterprise Communities* (1994).

TABLE 8.2 Regional Distribution of Model Cities and EZEC Designations

	Eastern[a]	Midwestern[b]	Remaining[c]	Total
Model cities	52 (35%)	32 (21%)	66 (44%)	150 (100%)
Urban EZECs	18 (28%)	12 (18%)	35 (54%)	65 (100%)

SOURCE: Frieden and Kaplan (1975); *News Release: Statement of Henry Cisneros on Announcement of Empowerment Zones and Enterprise Communities* (1994).
NOTE: Numbers in parentheses represent percentages.
a. Region defined as Eastern states: Maine, Massachusetts, Rhode Island, Connecticut, Vermont, New Hampshire, New York, New Jersey, Pennsylvania, Delaware, Maryland, District of Columbia.
b. Region defined as Midwestern states: Ohio, Indiana, Illinois, Wisconsin, Michigan, Minnesota, Iowa, Kansas, Missouri.
c. Region defined as remaining states: Virginia, West Virginia, North Carolina, South Carolina, Georgia, Florida, Tennessee, Kentucky, Alabama, Mississippi, Louisiana, Arkansas, Oklahoma, Texas, Arizona, New Mexico, Colorado, Utah, Wyoming, North Dakota, South Dakota, Nebraska, Montana, Idaho, California, Nevada, Oregon, Washington, Hawaii, Alaska.

Geographically, 18 (28%) urban EZEC designations were made in eastern states, 12 (18%) were made in midwestern states, and 35 (54%) were in the southern or western states usually considered a part of the nation's Sunbelt. Compared to award designations under the earlier Model Cities program, the urban EZEC designations were less favorable to eastern and midwestern communities and more favorable to those in the Sunbelt (see Table 8.2; Figures 8.1 and 8.2). During the approximately 30 years separating the two federal urban programs, HUD appears to have shown greater recognition of the issues of urban poverty and mob violence in southern and western cities and less in the older eastern and Midwestern Frostbelt urban centers. The growing strength in the House of Representatives among non-Frostbelt members helps account for this shift. And of course, it cannot be overlooked that HUD Secretary Henry Cisneros is himself from the Sunbelt. Perhaps not coincidentally, San Antonio, of which the secretary was once mayor, was among the cities receiving an enterprise community designation.

Each urban empowerment zone designee will receive a total of $100 million in social service block grants and tax relief for zone businesses. Of the two supplemental empowerment zones, Los Angeles will receive $125 million and Cleveland will receive $90 million. Among the urban enterprise community designees, each will receive $3 million. Among four enhanced enterprise communities, a total of $25 million will be awarded.

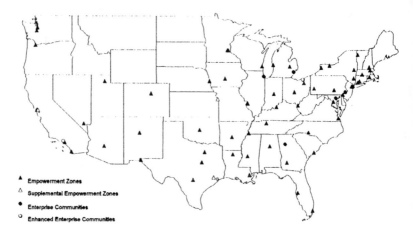

Figure 8.1. Distribution of Empowerment Zones and Enterprise
Communities Designations, 1994

SOURCE: *News Release: Statement of Henry Cisneros on Announcement of Empowerment Zones and Enterprise Communities* (1994); Map done by Ailsa Pratt, FAU-DURP, GIS Lab, 1995.

Figure 8.2. Distribution of Model Cities Designations, 1967-1968

SOURCE: Data from Freiden and Kaplan (1974); Map done by Ailsa Pratt, FAU-DURP, GIS Lab, 1995.

Twenty-seven years elapsed between the Model Cities and EZEC legislative campaigns. Both measures arose as historic congressional efforts to respond to major incidents of urban mob violence and the conditions of poverty underlying these events. Had the urban EZEC program turned out to be little more than what one enterprise zones advocate termed "a Model Cities throwback" (Cowden, 1995)? In the succeeding chapters, I will explore the shifts that occurred in Washington from the 1960s to the 1990s in problem definition, political reaction, and policy design directed at urban unrest.

Notes

1. A few weeks prior to the Los Angeles disorders, Kerry had called on the Senate to face squarely the issues of crime, social needs, and racism. His impassioned plea had been undergirded by a degree of candor about race and crime unusual on Capitol Hill. It had also emphasized the need for greater individual initiative and responsibility among the poor.

2. Although pro- and antiurban forces lent their voices to the debate on emergency assistance, the opportunity to pick up McConnell's rural plea was too much for Tom Daschle (D-SD) to resist. He introduced another amendment (1840) to H.R. 5132 calling on the president to designate as disaster areas farming communities suffering crop losses in the early 1990s. He was soon joined by Dale Bumpers (D-AK), who proposed additional money for small business loans in rural and urban disaster areas ("Federal Assistance," 1992). Few voices that day picked up the rural rallying cry.

3. By the end of the summer, it would become apparent again that the distance between good intentions and good outcomes is sometimes inversely proportional to the time it takes Congress to pass legislation. For the summer jobs section of the Urban Aid bill, the lack of careful thought and planning—admittedly necessary to piggyback the program on emergency legislation for Chicago and Los Angeles—left many cities ill prepared to contact employers, create jobs for needy youths, and publicize opportunities. As a result, in some cases jobs went unfilled because teenagers were unaware of their existence. In others, thousands of youths lined up for work when only a fraction of the needed jobs were available. For the lucky few, the program proved a success ("Conference Report," 1992, p. S17061).

4. Perhaps a bit disingenuously, Campbell indicated that over the 16 years of its life, the earlier riot reinsurance program "paid for itself"; in fact, he said, it "resulted in a net gain for the Treasury" ("Riot Reinsurance Act of 1992," 1992, p. E1935). What he failed to mention is that after 1968, catastrophic urban mob violence occurred only a few times. Had the program been enacted in 1964 or had large-scale rioting continued into the 1970s, the riot reinsurance program might have gone into the red. Then the Treasury would have had to subsidize the losses.

5. Like the Model Cities Bill, this proposal would spread program participation more broadly, enhancing its chances of enactment. But, of course, the more cities that participate, the fewer the resources per city and the less each neighborhood can achieve. Lloyd Bentsen uttered the same warning a few weeks later ("Enterprise Zones," 1992).

6. Riegle also introduced letters of support for the amendment from the U.S. Conference of Mayors, the National Congress for Community Economic Development, the American Federation of State, County and Municipal Employees, the AFL-CIO, the Center on Budget and Policy Priorities, the American Institute of Architects, and the National Association of Counties ("Tax Enterprise Zones Act," 1992, pp. S15027-S15028).

7. The Bradley initiative included a bill to train neighborhood youths to help repair and rehabilitate neighborhood facilities. It also proposed to keep public schools open after classes so that children could remain off the streets and engage in organized recreation and homework. Perhaps in deference to former President Carter's concerns, another bill would have subsidized community organizations to help create community banks. These institutions would help neighborhoods finance housing and job development, as well as other needed community assets. Other Bradley initiatives would assist city residents to become entrepreneurs, expand community policing, provide housing and child care for unwed mothers, assist urban workers to commute to suburban jobs, and set up tax-free savings accounts to help low-income families purchase a house, receive an education, start a business, or help pay for retirement ("Community Capital Partnership Act of 1993," 1993, pp. S5290-S5306). The eight pieces of legislation would be funded for a total of $1.45 billion per year over 5 years.

8. Enterprise zone bills had been passed by the Senate in 1982, 1984, and 1992. Three presidents had supported various zone measures; by 1992, 56 bills had appeared in Congress and 36 states had enacted state enterprise zone programs ("Sense of the Senate Resolution," 1993). In 1993, an additional 15 zone bills were introduced on Capitol Hill (Hornbeck, 1994).

9

Urban Poverty, Interracial Mob Violence, and Federal Reaction

The Problem and Political Contexts

In this chapter, I reflect on congressional response to urban poverty and interracial mob violence from the mid-1960s through the early 1990s. As Chapter 1 points out, there are three contexts to consider:

- The shifting nature of urban poverty and interracial mob violence since the 1960s (the problem context)
- The changing political environment within which these issues were debated and framed (the political context)
- The altered structure and content of federal programmatic responses to poverty and mob violence (the policy context)

This chapter addresses both the problem and political contexts. Chapter 10 covers the policy context.

The Problem Context

The explosion of mob violence in South-Central Los Angeles is now 4 years behind us. As I have shown, this catastrophe was only the most recent in a long history of such outbursts. But it was followed by the first successful attempt by Congress in a quarter century to enact new programmatic responses to urban mob violence. Was George Bush or Bill Clinton merely responding to the same set of problems that Lyndon Johnson faced in an earlier era? Have the conditions that contribute to urban poverty, particularly to minority urban poverty, remained unchanged since the 1960s? What of the character of mob violence itself? Was the Los Angeles disaster essentially a carbon copy of predecessors in Watts, Harlem, Detroit, and a host of other cities? Or do we have a new understanding of the conditions under which such violence occurs?

Nature of Urban Poverty in the 1990s

In the approximately three decades since the Model Cities bill was taken up, the nation has witnessed a profound transformation in the international economic order. Perhaps it belabors the obvious to note that neither our cities nor the nation nor the world are very much as they were 30 years ago. To understand how they have changed is to define some of the critical parameters within which the nature of urban life itself must be altered if meaningful futures are possible for those who suffer from poverty, discrimination, and debilitating living conditions. In the minds of many observers, the necessary departure point is the onset of the global economy and its handmaiden, the postindustrial society (Beauregard, 1989; Knight & Gappert, 1989).

As the third millennium approaches, the international economy is facing a restructuring in which the individual skills, manufacturing processes, and attainable levels of efficiency in goods production and distribution that were acceptable 30 years ago are now obsolete (Noyelle & Stanback, 1983; Stanback, Bearse, Noyelle, & Karasek, 1981). In the years since the 1965 Watts riot, capital has flowed to third world countries at an unprecedented rate, with the result that seemingly countless American factories, mines, mills, shipyards, and other industrial facilities have closed down or diminished their workforces (Kasarda, 1983, 1989). Of those that remain in business, many now contract with international firms (or have established their own branches) to supply needed parts, products, and services in places in which wage rates and benefits are lower than in the United States. On a smaller geographical scale, business relocation has also been occurring within the continental United States as many of the old Frostbelt plants have been closed or downsized and their jobs relocated to Sunbelt states (Hill & Negrey, 1987; Koritz, 1991; Sawers & Tabb, 1984). Trade policies such as the North American Free Trade Agreement guarantee that this process will continue, at least in the short run, with employers reestablishing themselves along (and on both sides) of our international boundaries. Because much of the nation's production historically has been located in or near central cities, the effects of these measures on the urban poor and working class have been felt disproportionately (Kasarda, 1993).

Meanwhile, as manufacturing employment declines, the services sector continues to grow (Hicks & Rees, 1993; Stanback et al., 1981). Essentially two types of jobs are offered: in basic services, such as fast food, clerical, housekeeping, personal care, and retail employment, and in advanced services, such as professional, technical, sales, and executive positions. Basic services often offer wages and benefits that are at best barely sufficient to support a family. Higher-paying advanced services, on the other hand, generally require postsecondary education. As competition rises for fewer manufacturing jobs and for lower-paying basic services jobs, lower-class whites generally compete poorly; minorities, especially African Americans, have even less success (Turner, Fix, & Struyk, 1991; Wilson, 1980, 1987).

Many cities are suffering diminished economic viability and are not likely to reverse their fortunes in the foreseeable future. Cities such as Buffalo, Gary, Camden, Detroit, St. Louis, and Cleveland no longer serve the primary commercial and industrial functions that originally brought them to economic power over the first half of the 20th century (Peterson, 1985). Many American cities were originally settled as maritime or rail traffic centers for the shipment and transshipment of goods. Most of these also became great manufacturing centers with the rise of the industrial revolution during the 19th and early 20th centuries (Kantor & David, 1988; Peterson, 1985; Wade, 1959). In the past 20 or 30 years, some cities have made the transition to an economy in which advanced services play a key role. Cities such as San Diego, Miami, San Francisco, Seattle, Minneapolis, and Atlanta—none of which were ever great centers of heavy manufacturing—have been able to adapt to the new world order because they have good accessibility to national and international markets for their services; in addition, the strength and diversity of their educational and research institutions, the skills and educational levels of their workforces, and their cultural, artistic, and environmental attractions have helped them acclimate to the postindustrial era (Stanback & Noyelle, 1982).

American cities have competed for economic activity with other cities and with their own suburbs for a long time. More than in the past, however, they must now compete internationally with other cities and regions for trade and investment. No longer able to dominate markets in steel or automobiles, for example, they compete for dwindling market shares (Glickman, 1983). The result is that there is simply not enough economic activity to go around. Instead of carrying on extensive trade with other U.S. regions or foreign nations, many of these places are forced to turn their economics inward. They have fewer industries and offices exporting goods and services outside their immediate region than was the case 30 or more years ago. Consequently, the trade of goods and services tends to be more localized, thus attracting less capital to city economies from outside the region. In addition, many businesses that remain viable nationally and internationally have relocated to the suburbs of these cities, where their tax revenues, jobs, and income are lost to urban governments and city residents. The Detroit region, for example, is still a major exporter of goods and services nationally and worldwide. But now most of the firms are located to the west and north of the city, in the thriving suburbs. Although recent gains in sales of American automobiles have raised hopes, many companies remaining in the beleaguered city of Detroit struggle to capture national and international market share (Bennet, 1994). Sadly, many old industrial centers have become redundant (Darden, Hill, Thomas, & Thomas, 1987).

Many of these cities now have larger populations than can be sustained by the available incumbent employment base, given prevailing education and skill requirements. Because urban and metropolitan economies cannot support large populations, civil service payrolls, and public services delivery merely by carrying on business activity within themselves, they must compete successfully to attract capital from other regions and other nations. As their ability to do so declines over time, those who live within their borders—especially those who are underprepared and discriminated against—are less and less able to achieve economic self-sufficiency.

In response to these views about the nature of economic restructuring, a series of counterproposals share the premise that local, state, national, and international political and governmental policies and actions help influence and shape the conditions driving investment, employment, income, and taxation (Wong, 1990). Citing examples of various local "growth machines" (Smith & Feagin, 1987; Wong, 1990) and "urban regimes" (Fainstein, Fainstein, Hill, Judd, & Smith, 1983), the counterproposals contend that municipal governments retain more control over their economic fortunes than theories about the market logic of capitalism would suggest (Logan & Swanstrom, 1990; Reese, 1993). In context with national policies affecting economic growth, some local governments are able to counterbalance potentially harmful trends such as corporate urban-to-suburban relocation, payroll downsizing, disinvestment, and housing loss. Private sector economic decisions do not always take place in isolation from the forces of human agency as represented by the state. Neither are they immune to the variabilities imposed by different customs, language, and culture (Logan & Swanstrom, 1990). Thus, economic restructuring does not carry with it immutable nor universal implications with regard to the future of cities.

Although observers may disagree over the particulars of such theories, that public policies can and do intervene in market forces to alter or redirect social and economic outcomes is accepted as a given. Rather, it is the viable range of choices left open to the state that has been compromised by globalization, deindustrialization, and economic restructuring. As a result, the federal government, as well as state and local governments, is forced in the 1990s to respond to issues of mob violence and poverty in a more severely constrained environment than in the 1960s. By no means does this fact exempt polities from the obligations of contributing to human welfare. But it does command the rethinking of policy bromides in both strategic and tactical terms.

Shifts in national and international economic conditions have been a major force in altering the context within which the problems of urban mob violence and poverty have been understood. Also contributing to the problem, however, is the nature of urban poverty itself. More than ever before, African Americans must now compete with other minority groups, primarily Asian, Latino, and Chicano, for jobs, housing, and government assistance. Meanwhile, they must also contend with poor whites, their primary source of competition in the 1960s. But fewer whites live in the larger central cities, so tensions between and among minority groups are now more common. (It is perhaps no accident that the two most recent outbreaks of urban mob violence in the United States occurred in Miami and Los Angeles, both of which have among the highest increases in the United States of new immigrants from Hispanic cultures.) Moreover, as the decline of manufacturing has disproportionately affected cities, new employment opportunities arise more commonly in the suburbs now. This "spatial mismatch" means that inner-city workers are at a disadvantage compared to their suburban counterparts because they lack the accessibility to those jobs (Ihlanfeldt, 1994; Kain, 1968). Coupled with increasing robotics and automation, which reduce the number of workers needed to produce goods and services (and usually raise the skill levels required), the result is that unemployment and underemployment have risen in central cities.

Poor African American communities have changed in other ways as well. The success of open housing laws and the rising number of African American college graduates contributed to the outmigration to the suburbs of thousands of moderate- to middle-income African American households. Left behind were mostly low- and moderate-income people (in what used to be called the ghettos). Their inner-city neighborhoods are now more completely isolated from the American mainstream than ever before (Massey & Denton, 1993). With fewer middle-class role models such as lawyers, police, firefighters, construction workers, and small business owners, poor households have a greater tendency to look to others of similar circumstances after which to pattern their behavior. Consequently, the subtle and not-so-subtle forms of social control imposed over low-income households by those more fortunate no longer operate as they once did (Wilson, 1987). Instead, many teenage black girls feel little guilt or remorse at bearing children outside of marriage. Many teenage black boys have little hesitancy about fathering children; far too many look to drug dealers and criminals for inspiration in carrying out their own day-to-day activities. Increased drug trade and substance abuse alone have brought profound changes in urban poor neighborhoods. To point these matters out is not to indict a generation of youths. On the contrary, it confirms the admirable fortitude and dignity of the many who manage to resist these temptations.

Nevertheless, these conditions are accompanied by higher proportions of single-headed households, the vast majority of which are headed by women. In many of these households, the mother has borne children with two, three, or more partners. Extended families, friends, and neighbors help care for the children in many cases, but the quality of housing and neighborhoods, the adequacy of diet and nutrition, and the effectiveness of public schools available to these families often are lower today than in earlier decades.

As a consequence of these and related conditions, the public and nonprofit sectors are asked to act as surrogates for the kinds of institutions and family and peer social structures that once were common in poor neighborhoods. Local, state, and federal government agencies, public schools, police departments, hospitals, and charitable institutions attempt to reweave these frayed threads; in many cases, they are overwhelmed by the demands. To be sure, single parent families are hardly unique to urban or poor neighborhoods; increases in divorce, separation, and out-of-wedlock birth rates have left their mark on middle-class and suburban communities as well. But the extent of poverty, hopelessness, crime, drugs, and other conditions in urban settings has created a far more problematic culture than exists in the suburbs. One result among many inhabitants of such neighborhoods is the perception of a severely circumscribed *opportunity structure* (Galster, 1993).

In public school systems, for example, teachers, administrators, bus drivers, and hall monitors are forced to play the role of parents, especially fathers, trying to enforce a degree of discipline among poor and minority children. Many youths carry weapons, witness drug deals, threaten and assault (or are threatened or assaulted by) others, fail to attend school, father children, and grow up in the absence of a father. Certainly, none of these tragedies is unique to the 1990s; they were in evidence in the 1960s. It is their magnitude and the diminished presence of counterbalancing forces such as neighborhood churches, social institutions, and stable families that account for most of the differences in the communities of the 1990s.

For these reasons, federal programs such as Model Cities are no longer appropriate for dealing effectively with urban poverty. Despite the multidimensional characteristics of Model Cities, at its heart was the notion that the poor simply need a temporary helping hand. With some job training, remedial education, better health care, food assistance, after-school recreation—with a host of services—the poor would "get back on their feet." This bootstraps mentality pervaded much of the War on Poverty legislation. It was a reasonable assumption in part because the interpersonal and social infrastructure described above was still much in evidence. Unskilled and semiskilled jobs were still available in many cities. Drugs had not so thoroughly pervaded community life as to threaten an entire generation. And attitudes about personal responsibility were more prevalent. Today, however, policymakers can no longer assume that a temporary helping hand is enough (*Confronting the Nation's Urban Crisis,* 1992). I will return to this issue in the next chapter.[1]

◦ Nature of Mob Violence in the 1990s

When the earliest major incidences of urban interracial mob violence appeared in the 20th century, they were based primarily on tensions between southern blacks who had migrated northward and working class white incumbents who felt threatened by competition for jobs, housing, and community facilities. Not infrequently, groups of whites attacked blacks in downtowns, black neighborhoods, or public places. Blacks reciprocated occasionally, but more commonly they retreated to their homes in the face of vastly superior numbers of white marauders. Generally, civil authorities attempted to prevent contact between the races, sometimes with favoritism toward whites.

After African Americans had established their own urban communities and their numbers reached a critical mass, a second generation of mob violence appeared. By the mid-1960s, these events were precipitated almost exclusively by conflict between one or two blacks (usually males) and symbols of white authority such as employers, merchants, or civil officials. The precipitating event was often the arrest or attempted arrest of a black youth for shoplifting, vagrancy, drunkenness, vandalism, a traffic infraction, or other minor crimes. These incidents frequently happened during hot summer weather, when tempers were short. With few exceptions, the resulting looting, arson, injuries, and deaths occurr near minority neighborhoods.

Although in the tradition of second generation urban interracial mob violence, the 1992 Los Angeles riots departed in at least two significant ways: police brutality was visually documented and crowd reaction was delayed.

Visually Documented Police Brutality

Violence in Los Angeles resulted not simply in response to another incident between an African American male and white police; it happened because, apparently for the first time, what appeared to be indisputable proof of brutality in the treatment of blacks by white police became available to the public. Unlike countless confrontations between police and white authorities in earlier riots, the savage beating of Rodney King was witnessed by millions. Seeing is believing, and few who viewed the videotape could deny the unfairness. Prior to the Rodney King episode, charges of police brutality were susceptible to the perceived veracity of the witnesses. Charges by African Americans could be easily dismissed by whites as due to crowd hysteria, youthful misjudgment, willful distortions, or the effects of alcohol or drugs. But the Rodney King videotape raised uncertainties about the possibility that some—perhaps many—past claims among African Americans of unfair treatment had had a basis in fact. In this respect, it played much the same role as the filmed beatings and hosings of black civil rights demonstrators did in southern communities during the 1960s. Just as sympathetic whites then saw with their own eyes the brutality suffered by African Americans, so did the Rodney King videotape reveal the viciousness attainable in police-citizen confrontations in the 1990s. But whereas the earlier events were underscored by long traditions of institutionalized southern racism, Rodney King was supposed to live in a time and place when such institutions had been eliminated nationwide. Conveyed to millions of viewers was the message that racism and brutality can and still do occur, at least at the hands of some law enforcement officers. As such, the veracity of claims by black and other minority victims of police violence may take on new credence in the future.

Delayed Crowd Reaction

The Los Angeles riots of 1992 were unusual not only because police brutality was visually documented but also because crowd reaction to the brutality was delayed for several months until the trial decision was announced. In this, the tragedy shared a pattern similar to that of the 1980, 1982, and 1989 Miami riots. Each event involved white police officers accused of assaulting one or more African Americans during arrest procedures. In the 1980 and 1989 incidents, one or more of the arrestees was killed. In 1980, 1989, and 1992, a trial occurred; in each, the accused police officers were ultimately acquitted by the jury. In 1980 and 1992, serious and extensive rioting followed the court verdicts. These two tragedies departed from earlier patterns of mob violence in that they occurred not in the heat of the moment, but many months thereafter. Previously, rioting was often attributed to spontaneous crowd reaction, either in witness to police-victim confrontations or shortly thereafter in response to rumors that followed.

That widespread violence did not proceed immediately after the 1980 Miami beating death was, in itself, unusual by historical dimensions of interracial mob violence. But there was no recorded visual evidence of the event. In the case of Rodney King, however, the shocking scenes of police officers repeatedly pounding and kicking their seemingly unresisting victim were viewed by millions within a short time after the incident. The King incident thus became the first in history in which visually recorded evidence helped verify and substantiate the indignation and anger of a rioting crowd. For some observers, it conveyed indisputable proof of the kinds of brutality to which minority victims of police arrests are sometimes subjected. For others, the videotape opened nagging uncertainties about arrest circumstances preceding earlier riots and about how the American public should judge such events in the future.

In spite of these unprecedented circumstances, mob violence did not immediately break out in Los Angeles when the videotape was telecast. Instead, the African American community in South-Central deferred action until the jury's verdict was rendered. The disorders that followed, unlike previous mob violence, appear to have proceeded as an amalgam of emotion growing out of the ostensible injustice of the court's decision and the precipitating event. Rioters had had many months to absorb and confront the beating incident. Yet even after a cooling-off period, the most deadly and most destructive urban riot in modern times resulted. The 1980 Miami and 1992 Los Angeles disasters demonstrate that urban interracial mob violence erupts not solely in response to immediate antagonizing events; it can happen even in the aftermath of the opportunity for reflection, investigation, and thoughtful interaction between races and classes. Even with a year to cool inflamed passions, the propensity to strike out was too powerful to be denied.

The Los Angeles tragedy, partially foretold by earlier events in Miami, constitutes a new chapter in the history of urban interracial mob violence in America. It eroded some classic explanations of these events; for example, mob violence does not happen solely in the heat of summer nor always in spontaneous outbursts propelled by rumors or by direct witness to police-citizen confrontations. Even with months of time for the courts, the media, churches, and other institutions to intervene, to calm, to establish facts, and to counsel prudence, the threat of disastrous outbreaks of mob violence persists. Furthermore, the home videocamera adds a new dimension in the human capacity to delay witness to events and verify their circumstances. Just as innovations such as the automobile, the television, and the computer have been "democratized" through mass production, so is the home videocamera becoming ubiquitous. The result, as numerous broadcast network documentary programs are now demonstrating, is that criminal behavior and police-citizen confrontations are more likely to be captured on videotape. Put another way, human experience has risen to a new level of replicability and portability. The outcome will likely be more examples of Rodney King-type video documentation. Accompanying them will be a new capacity for the media, the criminal justice system, and minorities who charge discriminatory treatment to challenge, if not always clarify, the truth about events.

The Political Context

I have argued that the context encompassing urban interracial mob violence and poverty has undergone significant changes since the 1960s. But it is clear that the political context in Congress within which these issues are debated also has shifted substantially. This contrast is demonstrated in my earlier description of the terms of discourse in legislative periods on Capitol Hill in response to urban mob violence in the mid-1960s and the early 1990s. In particular, seven conditions emerge that frame the altered nature of political discourse. The first two encompass structural and resource issues in Congress, whereas the latter five involve the more elusive evolution of ideas and collective attitudes about poverty, human rights, race, violence, and personal responsibility:

- Regional, partisan, and racial shifts in congressional membership
- Competing budgetary demands and fiscal constraints
- Desire to ease poverty and the liabilities of precedent
- Loss of the moral climate of the civil rights movement
- Decline in racially divisive discourse
- The violence conundrum
- Metaphorical revisionism

Regional, Partisan, and Racial Shifts in Congressional Membership

One condition that helps explain the political context within which riot-responsive measures have been debated is the changing composition of Congress itself. By the time of the 1992 Los Angeles riot, several forces had intertwined to produce a legislative body whose membership differed substantially from that of the mid- and late 1960s. Few in Congress in the wake of the Los Angeles riots were members of that body when Harlem, Watts, or Detroit exploded in violence in the mid-1960s. As Table 9.1 shows, only six senators who were members when the Model Cities bill was approved were still members when the enterprise zones-empowerment communities (EZEC) legislation was enacted.

Among House members, only 23 (5%) were in office when both measures were voted on. In both chambers, most of these enduring figures were Democrats. Although it is hardly surprising that the composition of Congress turned over so drastically in 27 years, the comparatively small number of War on Poverty era legislators indicates how little hindsight Congress had to draw on in formulating its newest response to urban interracial mob violence. Few on Capitol Hill had ever grappled with the seemingly intractable political issues such events produce. It is safe to

TABLE 9.1 Members of the 89th Congress Who Were Members in the 103rd
              Congress

Senate	*House*
Daniel Inouye (D-HI)	Don Edwards (D-CA)
Robert J. Dole (R-KS)	George E. Brown, Jr. (D-CA)
Edward M. Kennedy (D-MA)	Sam M. Gibbons (D-FL)
Claiborne Pell (D-RI)	Patsy T. Mink (D-HW)
James Strom Thurmond (R-SC)	Frank Annunzio (D-IL)
Robert C. Byrd (D-WV)	Daniel D. Rostenkowski (D-IL)
	Sidney R. Yates (D-IL)
	Lee H. Hamilton (D-IN)
	Andrew Jacobs, Jr. (D-IA)
	Neal Smith (D-IA)
	William H. Natcher (D-KY)
	John Conyers, Jr. (D-MI)
	William D. Ford (D-MI)
	John D. Dingell (D-MI)
	Jamie L. Whitten (D-MS)
	Joseph M. McDade (R-PA)
	James H. Quillen (R-TN)
	Jack B. Brooks (D-TX)
	James J. Pickle (D-TX)
	Eligio de la Garza (D-TX)
	Henry B. Gonzalez (D-TX)
	Thomas S. Foley (D-WA)
	Robert H. Michel (R-IL)
Total = 6 (6%)	Total = 23 (5%)

assume that fewer still had any substantive knowledge of the Model Cities
or War on Poverty legislative or programmatic histories on which new
responses to urban mob violence might be framed. As a result, congressional response to urban mob violence occurred under circumstances quite
unlike the vast majority of issues encountered on Capitol Hill. Rather than
a continuous concern, urban mob violence is sporadic and highly discon-
tinuous. There is no congressional committee or federal agency charged
with oversight on the issue. In the absence of an ongoing institutional

memory, effective monitoring of such events is lacking. The very characteristics of randomness and infrequency that typify mob violence render
the federal government unprepared and unschooled to grapple with a
response when outbursts occur.

What factors account for the vast turnover of Congress in the period
between Model Cities and EZECs? Obviously, many members retired or
were defeated for reelection. But beneath these dynamics lies the powerful
effect of interregional population shifts in the United States. Throughout
the 1970s and 1980s, large numbers of households were migrating to the
so-called Sunbelt states (Frey, 1993). There is no universal agreement on
which states are included in the Sunbelt, but most definitions include all
the southeastern coastal states from Virginia to Florida, the Gulf coastal
states from Alabama to Texas, the southwestern states of New Mexico and
Arizona, and California. The Pacific coastal states of Oregon and Washington are sometimes included. Although some of the Sunbelt's growth
resulted from births and international immigration, much was due to
migration from the regions of the nation losing population. The so-called
Frostbelt includes "Rustbelt" states in New England, the Middle Atlantic,
and the Midwest, as well as the Plains states. The Frostbelt encompasses
most of the nation's older, larger industrial, railroad, and port cities.

When Congress was enacting War on Poverty legislation in the late
1960s, the prevailing demographic patterns were rural-to-urban migration
from southern agricultural states to the urban North and outward migration
from central cities to the metropolitan suburbs. The largely westerly and
southerly migration patterns from the Frostbelt to the Sunbelt were evident
in only a few states such as Texas, Florida, and California. By the 1980s,
however, a powerful shift was under way in the nation, led by retirees
seeking sunshine and mild weather and workers after new job opportunities
in defense, aerospace, computers, and other high-tech industries.

In response, it became necessary to redraw congressional district lines
to reflect emerging population redistribution better. Between 1966, when
the Model Cities bill was enacted by the 89th Congress, 2nd Session, and
1993, when the EZECs measure was adopted by the 103rd Congress, 1st
Session, a total of 46 congressional districts were redistributed. As Table
9.2 shows, 19 states lost and 12 states gained one or more members of
Congress in their delegation. All the losers were located in the Frostbelt,
whereas all the gainers were in the Sunbelt. Among the largest shifts, New

TABLE 9.2 States Whose House Delegations Increased or Decreased in Size Between the 89th and 103rd Congresses

Gains (12)		*Losses (19)*	
Arizona	3	Alabama	1
California	14	Illinois	4
Colorado	2	Indiana	1
Florida	11	Iowa	2
Georgia	1	Kansas	1
Nevada	1	Kentucky	1
North Carolina	2	Louisiana	1
Oregon	1	Massachusetts	2
Texas	7	Michigan	3
Utah	1	Missouri	1
Virginia	1	Montana	1
Washington	2	New Jersey	2
		New York	10
		North Dakota	1
		Ohio	5
		Pennsylvania	6
		South Dakota	1
		West Virginia	2
		Wisconsin	1
Total	46	Total	46

York's House delegation declined by 10 members, Pennsylvania's by 6, Ohio's by 5, and Illinois's by 4. Meanwhile, California's congressional delegation rose by 14, Florida's by 11, and Texas's by 7. The rising strength of the Sunbelt and the declining presence of the Frostbelt during the 1970s and 1980s did much to undermine support in Congress for initiatives that could disproportionately benefit the old manufacturing cities of the Northeast and the Midwest.

Congressional redistricting was not the only influence that accounted for the new composition of members on Capitol Hill. Brought about largely by the decline in representation among Frostbelt urban states, the growing Sunbelt delegations increased the presence of Republicans. As Table 9.3 shows, the Senate in the 89th Congress was composed of 68 Democrats

TABLE 9.3 Changes in Congressional Membership by Party, 89th and 103rd
Congresses

	89th Congress		
	Democrats	*Republicans*	*Total*
Senate	68	32	100
House	295	140	435
	103rd Congress[a]		
	Democrats	*Republicans*	*Total*
Senate	56	44	100
House	258	176	435[b]

a. 112 new representatives joined the House in January 1993: 64 were Democrats and 48 were Republicans.
b. In addition to Democrats and Republicans, one member is an Independent.

and 32 Republicans. In the 103rd Congress, Democratic strength had
slipped to 56 senators, whereas Republican might had risen to 44. Similarly, the earlier Congress had 295 Democrats and 140 Republicans in the
House; the 103rd contained 258 Democrats, 176 Republicans, and 1
Independent. Clearly, Democratic voting power had slipped significantly
since the days of Lyndon Johnson's domination of the Congress. By 1993,
Democrats no longer wield the influence they once did.

If the Democratic grip has slipped, the loss has not come entirely at
the expense of African American interests. Although blacks had a strong
stake in both the Model Cities and EZEC measures, the number of blacks
who were members of Congress rose over the intervening period. Table
9.4 shows that the 89th Congress contained only 5 black House members,
whereas the 102nd Congress had 26 blacks, an increase of more than 400%.
After the November 1992 elections, black House membership in the 103rd
Congress rose to 40. (The Senate had no black members in either Congress,
although Republican Edward Brook of Massachusetts served in the Senate
during the interim between these Congresses.) Sixteen of the 40 black
members were newly elected; 13 were from House districts redrawn
specifically to enhance black political power under authority of the 1965
Voting Rights Act.

In summary, Congress has seen a diminishing influence among the
traditional Democratic strongholds in the industrial cities of the Frostbelt

TABLE 9.4 African American Members of Congress, 89th, 102nd, and 103rd Congresses

89th Congress	
House	5
Senate	0
102nd Congress	
House	26
Senate	0
103rd Congress	
House	40
Senate	0

and a corresponding rise in power within emerging Republican centers of strength in the suburbs and in the Sunbelt states. With the rise in black House membership, traditional Democratic urban interests do not appear to have shifted substantially. Yet black and white Democrats do not by themselves compose so decisive a voting block as they once did on Capitol Hill. Moreover, minority constituencies have shifted as many African Americans have left cities for the suburbs and their places have been taken by Asians and Latinos. All these factors help to account for weakening political dedication in Congress to the problems of urban industrial cities and their high concentrations of poor citizens. Poor neighborhoods of the type targeted by Model Cities tend not to be as severely deteriorated physically or economically in the generally younger Sunbelt cities. Of course, Frostbelt and Sunbelt cities share common poverty problems of homelessness, unemployment, crime, and drug abuse, and urban problems of private disinvestment, deteriorating municipal tax bases, and the like. But with the possible exception of a few centers such as Los Angeles, many of these conditions are not as advanced in Sunbelt cities as in their Frostbelt counterparts (Bradbury, Downs, & Small, 1982).

Not surprisingly, as the regional, partisan, and racial makeup of Congress has changed, so has the nature of urban politics. As the development of a federal response to urban interracial mob violence unfolded during 1992 and 1993, it was apparent that the liberal Democratic urban voting block no longer carried the weight it once had. Bills sponsored by Edward Kennedy, Donald Riegle, and William Bradley, for example, had many of the earmarks of War on Poverty-type measures. The emphasis in most of these older-style programs was on increased federal spending in cities for various services to the poor. With the rise in Republican and Sunbelt members of Congress, however, interest has shifted to include more self-help programs centered on private investment incentives, job training, and job creation. Not surprisingly, the EZEC measures stressed these goals.

Competing Budgetary Demands
and Fiscal Constraints

Virtually no legislation before Congress escapes consideration within the confines of the budgetary process. This was true in 1966 and it was true in 1992 and 1993. The fiscal appetites of both Model Cities and EZECs were balanced among competing demands for the nation's limited resources. Thirty years ago, measures relating to civil rights and poverty competed with others addressing domestic needs such as farm price supports, highway construction, defense spending, public education, and Social Security. Since then, American society has discovered a plethora of groups who are disfranchised and discriminated against. Thus, minorities such as African Americans and the urban poor must compete with women's rights, gay rights, handicapped rights, animal rights, workers' rights, consumers' rights, abortion rights, and the like for programmatic public support. Even if none of these competing interests had emerged, however, there are more recognized minority groups than existed in the Great Society era. Then, African Americans received the lion's share of attention from Washington and the media. Now, African Americans must compete with other protected classes such as Asian Americans, Native Americans, and Hispanic Americans for public subsidies, regulatory protections, preferential college admissions, minority job and public contract set asides, and the like. Added to these challenges are the enormous costs of achieving compelling environmental objectives such as clean water, clean air, wetlands protection, habitat and species preservation, and removal, storage, or neutralization of toxic, hazardous, and radioactive waste. Because these environmental conditions threaten human life and health as well as the survival of other species, they appeal to a much larger share of the American public than issues pertaining to minorities or human rights issues. Consequently, traditional programs serving urban African Americans are now measured against a much broader array of budgetary demands than existed in the mid-1960s.

Compounding this situation in the 1990s is a seemingly uncontrollable national debt. War on Poverty programs such as Model Cities proceeded in sounder economic times. To be sure, the Vietnam War and defense spending posed constraints on congressional willingness to fund new urban and poverty initiatives. But American business generally prospered during the 1960s, and federal revenues rose accordingly. Although the nation was at peace in the post-Desert Storm period during which the EZEC legislation was considered, it was forced to contend with accumulated debt far in excess of that faced by Congress in Lyndon Johnson's day. In particular, expenditures on entitlement programs such as Social Security and Medicare were rising far faster than Congress had ever predicted. Even with the dissolution of the Soviet Union and the end of the Cold War, a decline in defense spending could not, by itself, offset these and other hefty outlays. Although official Washington will always disagree as to what constitutes excessive spending, there can be little doubt that the fiscal climate of the early 1990s was less forgiving of new funding commitments than it had been during the heyday of the Great Society.[2]

Desire to Ease Poverty
and the Liabilities of Precedent

Among the circumstances that resulted in ultimate congressional enactment of the Model Cities and EZEC legislation was the unquestioned commitment of several members to reduce the incidence of poverty, especially urban poverty, in America. For some, a constructive response to the violence in Watts and later in South-Central was an issue of moral and ethical responsibility. For others, poverty reduction could be justified on grounds of expediency. After all, crime control, substance abuse therapy, welfare, subsidized housing, community development, remedial education, and the other services associated with poverty cost American taxpayers billions of dollars. Pilfering, shoplifting, vandalism, robberies, arson, hijacking, and other crimes contribute to higher prices at the cash register. Furthermore, the existence of a large class of indigents in one of the wealthiest nations on earth is embarrassing to America's diplomatic image abroad. Whatever their motivations, some members of the House and Senate were committed to easing poverty as an end in itself. But linked to these beliefs was the assumption of some members that by mitigating unemployment, underemployment, poor health, boredom, unfit housing conditions, and the like, the United States could reduce the likelihood of further urban mob violence. Underlying this assumption was the notion that, precipitating incidents notwithstanding, human deprivation is the root cause of rioting. Traditionally, moderate-to-liberal Democrats have been more centrally identified with this posture than Republicans.

One is not surprised then, by the display of liberal Democratic sensibilities in the face of charges that the War on Poverty had failed. Both Attorney General William P. Barr and White House aide Marlin Fitzwater blamed conditions that contribute to rioting on misguided liberal Democratic policies of the past. Congressman Henry Gonzalez (D-TX) found Barr's comments "outrageous," "offensive," "reprehensible," and "despicable" ("Introduction of Legislation," 1992, p. H3080). Senator Daniel Moynihan (D-NY) labeled Fitzwater's words "a lie," "depraved," and a "slander" ("Setting a Firm Course," 1992, p. H3226) on Lyndon Johnson's memory. Others such as Congressman Jake Pickle (D-TX) offered a calmer, yet determined, defense of Johnson's policies. What accounted for the tender sensibilities of liberals on Capitol Hill?

When the Great Society was being legislatively and bureaucratically constructed in the mid- and late 1960s, the nation was embarking on its greatest period of social experimentation since the New Deal. Model Cities was among the most experimental of the many War on Poverty programs. Although Republican and conservative opponents insisted that such programs would not work, they could not prove their charges, for few programmatic precedents were available to mar the integrity of even the most exuberant claims. The nation's sheer paucity of experience with policies specifically designed to assist the urban poor limited the empirical grounding within which opponents could attempt to undermine new proposals. Coupled with the compelling immediacy of the civil rights movement and unrest in the cities, efforts to pass progressive new legislation benefited from the very uniqueness of the proposals themselves.

By the early 1990s, however, a vast legacy of War on Poverty experience had accumulated. Over the years, numerous studies by university scholars, government agencies, and foundations had documented the performance of many of these programs. Several people had testified before various committees on Capitol Hill. More important, the general perception in America was that the nation's poverty concentrations—the slums and ghettos of the 1960s—were still with us. In short, the War on Poverty had left many tracks behind and opponents could now charge that most of the footsteps led nowhere. Liberal Democrats could no longer capitalize on the unfamiliarity of the nation with programs designed to assist the poor. In the wake of almost 12 years of conservative-to-moderate Republicanism in the White House, charges of the failure of the War on Poverty cut close to the bone, especially for the declining legions of Great Society veterans such as Gonzalez, Moynihan, Pickle, Kennedy, and others. Now, for the crime of having tried, Democrats found themselves on the defensive.

As Ronald Reagan popularized supply-side themes of deregulation, privatization, and massive tax and spending cuts, Republicans ultimately found themselves in the position of having to respond to the continuing dilemmas of urban poverty. Whatever it was, it had to look significantly unlike anything that smacked of Democratic liberalism. Borrowing from the Thatcher government, Republicans launched their proposals for public housing devolution and enterprise zones in the 1980s. Now they had the advantage, pointing out that measures designed to further private property ownership, business development, and job creation were largely untried and represented a sharp departure from the tired legacies of the New Deal and Great Society. Individual self-interest, they insisted, would do more to deliver the nation's poor from poverty than liberal direct spending programs had ever done. Once again, there was little, if any, research to support or refute partisan claims when these measures first arose.

It appeared to matter little that, although no previous national program of enterprise zones had been enacted, several state-sponsored initiatives had been in existence since the mid-1980s. Even though the results of limited studies of these state programs were mixed, in the absence of federal tax and regulatory relief, enterprise zone supporters could argue that the jury was still out. Realizing that 1960s-style large-scale direct spending programs were unlikely to gain Republican or conservative Democratic support, most Democrats were willing to settle for an enterprise zones program as better than nothing at all. This factor, and the lack of a convincing precedent by which to evaluate claims for the enterprise zone concept, ultimately helped propel the EZEC measure to passage. In this respect, supporters of the enterprise zone concept in the early 1990s found themselves in a position not unlike Model Cities advocates in 1966.

Loss of the Moral Climate
of the Civil Rights Movement

If both Model Cities and EZECs were advanced in response to record-setting incidents of urban interracial mob violence, the discourse on EZECs did not have the advantage of the civil rights moral crusade waged by groups such as Martin Luther King, Jr.'s Southern Christian Leadership Conference. The civil rights movement was founded on the indisputable facts of de jure and de facto discrimination in voting practices, housing, jobs, and a host of other sectors of American life. For the most part, the divisions of right and wrong, morality and immorality around which Congress and the Johnson White House framed War on Poverty legislation were relatively clear.

To point out these conditions, however, is not to maintain that either Congress or the American public was unanimous in subscribing to them during the 1960s. There were bitter divisions in the South about the place of blacks in community life and in politics. Similarly, strong differences existed on issues such as the primacy of states' rights, the meaning of the Constitution, and the threat to public order posed by racial desegregation and equal opportunity policies. Moreover, these divisions were hardly unique to the South; to a lesser extent they existed in other regions of the nation as well. Nonetheless, few could deny the existence of de jure discrimination in the South or de facto discrimination in the nation as a whole. Although many disagreed about the extent or the causes, most would not disagree that blacks were disproportionately disadvantaged. For these reasons, the moral and ethical issues of the civil rights movement stood out in sharp relief.

Urban poverty and the conditions breeding mob violence, although more complex, could be associated with these injustices by liberals. Thus, it was not merely indolence, low education, or weak job skills that accounted for the plight of the urban poor. Bigoted landlords, discriminatory employee practices, and thoughtless public policies such as Urban Renewal also helped account for their suffering. The issues of civil rights and urban poverty were associated with one another.

By the early 1990s, however, although racial inequality and urban poverty persisted, the moral basis of rioting as a response could no longer be founded as effortlessly on blatant de jure practices such as segregated public facilities, denial of the right to vote, and discrimination in the awarding of public and private sector jobs. Racial and ethnic discrimination continue, but are considerably more likely to occur through more subtle practices. For the most part, when discrimination takes place today it is practiced sub rosa by some landlords, real estate agents, employers, and merchants, for example (Elmi & Mikelsons, 1991; Turner, 1992). As federal audits have shown, in some situations racial minorities are denied equal treatment due to wrongful intent; in many others, unequal treatment cannot be proven and unequal outcomes result from actions taken with no apparent or provable ill motivations. Given these circumstances, it is difficult to launch another nationwide moral crusade against these practices because it is often impossible to prove that individual instances of discrimination have occurred. It is even more difficult to demonstrate that such acts are part of an ongoing and consistent pattern or practice and not merely isolated occurrences of bias. In the former, federal policy intervention is usually merited (as it was in the 1960s); in the latter, litigation or other civil measures are the more likely courses of action.

For all the cruelty and injustice of the Rodney King beating incident, that event and the court's acquittal of his tormentors could not, by themselves, equate with the stark moral power and magnitude of Martin Luther King, Jr. Events associated with Rodney King have opened new grounds on which future incidents of mob violence might be understood. But these conditions are not likely to undergird future urban poverty initiatives with the magnitude enjoyed by Model Cities and other War on Poverty programs, associated as they were with national civil rights reform.

Decline in Racially Divisive Discourse

If the moral dimensions within which the EZEC legislation was debated in Congress were considerably diminished by comparison with those of Model Cities, the EZEC measure was also free from the explicit racism that clouded consideration of its predecessor. Concerns from some members of Congress over the threat of mandatory school busing or racial desegregation in housing under Model Cities became the foundation for efforts to water down, if not defeat altogether, progressive efforts at reform. Dixiecrats such as Joseph D. Waggoner (D-LA), who pointed out that in his town the Howard Johnson's offered only vanilla, provided not-so-subtle reminders that there was little tolerance for federal programs that aspired to advance the cause of blacks. Conservative southern Republicans such as Prentiss Walker of Mississippi chastised Model Cities because it threatened to make "the many races created by God into one race" (*Congressional Record,* 1966b, p. 27013). Just beneath the surface of his comments lay a legacy of southern antimiscegenation laws, exploiting white fears of interracial sexual relations. Republican Congressman Albert W. Watson of South Carolina forsaw Model Cities money being used as an "economic pistol" (*Congressional Record,* 1966b, p. 26941) to the heads of city officials to entice further residential and school integration.

Racial animosities were not limited to Southerners. In Model Cities, Bronx Republican Congressman Paul Fino saw "zebra-colored housing and education guidelines" (*Congressional Record,* 1966b, p. 26922). Playing on white fears, he warned that the measure would become a gravy train and a bankroll for black power. Brooklyn Democratic Congressman Abraham J. Multer succeeded in amending the Model Cities measure to ensure that future participants would not be required to bus students to achieve racial desegregation in public schools as a condition of receiving continued aid.

By contrast, such overtly racial references were absent from the EZEC debates on Capitol Hill. Whatever biases lurked in the hearts of individual members of Congress, revealing them in public was no longer permissible. Even conservative southern Republicans such as Senator Phil Gramm of Texas steered clear of the race issue. Instead, reservations were based on other grounds, such as procedural or fiscal matters. Questions were raised about the propriety of attaching any new urban legislation to a post-riot emergency aid bill. Fiscal issues such as reducing funding for enterprise zones and Weed and Seed programs, rather than defeating the programs altogether, also gained attention. The fact that Republican President Bush and former Congressman and Department of Housing and Urban Development Secretary Kemp were behind the 1992 measure meant that even GOP conservatives could not stand in the way with impunity. In 1993, although Democratic President Clinton was behind his own version of enterprise zones, Republicans who had voted for previous bills could not easily reverse their position and oppose them. Thus, aside from whittling their budgets down, there was never serious resistance to the EZEC bills per se. Instead, where opposition occurred, it was directed at tax relief or rebudgeting issues associated with the omnibus bills to which the urban policy measures were attached.

The presence of the Congressional Black Caucus, in existence for more than a decade, diminished any propensities in the House or the Senate toward racially divisive rhetoric. In 1992, the Black Caucus numbered 25 members; by 1993 its ranks had increased to 40, due largely to redistricting in several southern states. Thus, for all the racial overtones associated with the Rodney King beating itself, the EZEC bills succeeded in demonstrating that the tenor of congressional discourse on urban policy had progressed remarkably since the War on Poverty era.

The Violence Conundrum

Both the Model Cities and EZEC measures were enacted not only as a response to urban poverty but also as an antidote to urban interracial mob violence. Within the violence rationale, though, two opposing views colored congressional consideration of each measure. On the one hand, some members of Congress during the mid-1960s warned of further urban rioting if the nation should fail to enact programs to address the issues allegedly at the root of outbreaks such as those in Chicago, Cleveland, and Philadelphia (i.e., the danger-of-violence argument). As early as mid-June 1966, for example, it was clear to Senator Muskie that in the eyes of the Johnson White House, rapid action was necessary to avert further Watts-style civil disorders in U.S. cities. Accordingly, Muskie did not fail to employ the danger-of-violence rationale in support of Model Cities (Haar, 1975). Senator Gaylord Nelson forsaw a nationwide crisis without further spending on social programs. Ohio Senator Stephen Young worried that more violence like that in Cleveland would erupt without congressional action. In August, Attorney General Katzenbach warned a New York audience that 30 to 40 cities could erupt in violence that summer. Johnson's own fears (Lemann, 1991) and the apocalyptic national mood (Bornet, 1983; Fogelson, 1969a; Frieden & Kaplan, 1975; Haar, 1975) were clear expressions of the crisis mentality in which the legislation—and some of its Great Society successors—was born.

Similarly, in the wake of the 1992 Los Angeles tragedy, some in Congress sought to ignite the torch of conditional violence. During floor debates on a post-riot emergency appropriation, Congressman James A. Traficant (D-OH) warned that several American cities were in danger of "blowing up in flames" ("American Cities in Danger," 1992, p. H3097). Congressman Major R. Owens (D-NY) feared "more death and destruction" unless Congress reestablished the war on poverty ("The Great Tax Conspiracy," 1992, p. H8132). Congressman Howard Wolpe (D-MI) saw the possibility that the nation could experience a "hell of racial conflict" and "racial polarization" ("Setting a Firm Course," 1992, p. H3234). Congressman Charles B. Rangel (D-NY) announced that, without the enterprise zone measure, American inner cities would "blow up" ("Conference Report," 1992, p. H11638). In the other chamber, Senator Donald Riegle feared "more Los Angeles in communities across the Nation" (*Congressional Record,* September 15, 1992, pp. S15027-S15028) unless the Senate acted positively. Senator Christopher J. Dodd (D-CT) forsaw "future disasters" like that in Los Angeles should the nation fail to enact new urban legislation ("Federal Assistance," 1992, p. S6977).

These warnings notwithstanding, the danger-of-violence rationale gradually lost credence on Capitol Hill in the weeks following the Los Angeles riot. With the threat of a clear-and-present danger apparently past, the danger-of-violence argument became untenable and its leverage shrunk. Still, anxieties about spreading riots appear to have had some effect in propelling the emergency appropriation forward in May. Smaller, much less damaging outbreaks of mob violence occurred in several cities parallel with the LA disorders and fed the fears of some members from urban states and districts. Barely a week after the Los Angeles disorders, Senator Kennedy, one of the few veterans of the Model Cities debate still in the Senate, recognized this mood on Capitol Hill. He quickly put forth a bill to spend an additional $5 billion on urban and welfare programs in cities across the nation. Sensing that sufficient senatorial endorsement was unlikely, he lost no time in cosponsoring a less ambitious bipartisan measure with Republican Senator Hatch. Ultimately, the Kennedy-Hatch amendment attached a half-billion dollars of general urban assistance to the emergency appropriations bill.[3]

By June, when H.R. 11, the enterprise zones bill, was under consideration, the danger-of-violence rationale had lost ground. As the Model Cities bill had 26 years before, H.R. 11 worked its way through Congress during the dog days of a Washington summer and into the early fall. But unlike with the previous proposal, no outbreaks of urban mob violence followed. Those on Capitol Hill who championed enterprise zones could not credibly advance the danger-of-violence argument in support of their position. Presaged by outbursts in Miami during the 1980s and smaller mob events in Washington in the early 1990s, record-setting rioting in Los Angeles proved insufficient to carry much influence by the time the measure was passed by the House in early July. As it plodded through the Senate in August and September as part of the Revenue Act of 1992, congressional attention was rooted in tax relief for constituents and was astray from any thoughts about an epidemic of urban violence. The following year, after President Bush failed to sign the Revenue Act and the Clinton administration was pressing the EZEC bill, there were no attempts on the House or Senate floor to rekindle the theme of imminent mob violence. The danger-of-violence rationale, so often posed by supporters of Model Cities and other War on Poverty measures, carried far less political weight when EZEC measures were taken up.

Just as the danger-of-violence argument gained only a small foothold in congressional floor debates on emergency appropriations, so did the second violence theme from the Model Cities era—rewarding the rioters—gain little support. When cities were exploding in the 1960s, a common response from conservatives was that to spend tax dollars to repair riot damage, reestablish businesses, rebuild housing, and provide social assistance programs to the poor was to reward those guilty of causing the mayhem. As a counter to the danger-of-violence rationale, the reward-the-rioters response insisted that further spending conveyed the message that violence begets compensation. As the list of War on Poverty programs expanded over the latter years of the Johnson administration, their total fiscal outlay (added to costs of preexisting urban programs such as Urban Renewal) gave opponents ammunition to argue that the United States was already spending too much (or enough) on cities and the poor. If these expenditures had failed to prevent the riots, how could more spending do so? Regardless of the merits of such claims, their popular appeal was not inconsiderable. In this context, new initiatives such as Model Cities risked portrayal at the hands of some Republicans and conservative Democrats as little more than compensation for those who carried out the violence.

Further complicating these charges was the often pyrotechnical rhetoric of black radicalism. As groups such as the Black Panthers organized in city after city, many became susceptible to the conclusion that riots were anything but spontaneous revolts against alleged injustices. Warnings by black militants that urban violence could reach white neighborhoods fueled conservative charges that additional spending would reward those responsible for mob actions. Congressman Fino's admonition that a "black power takeover" would tap into Model Cities funds in San Francisco, bringing rioting and "anarchy" was one expression of this view. Public opinion polls showed that a significant minority of whites—particularly among lower-income households—believed that blacks preferred rioting and violence.

Not surprisingly, the reward-the-rioters theme resurfaced in the wake of the Los Angeles disorders of 1992. The Bush White House lost little time in releasing a trial balloon that pointed out that the nation had already spent billions of dollars on urban problems with few successes. Implied was the message that, had the legacy of War on Poverty programs been effective, violence such as that in Los Angeles would never have erupted. In calling for a resurgence of family values, Bush signaled that his administration was not interested in mounting another era of massive federal spending for urban and poverty programs. At about this time, whether by commission or by carelessness, Attorney General Barr and White House aide Marlin Fitzwater delivered their failed War on Poverty volleys, reaping rebukes of seismic proportions from Democrats. With an election coming up in the fall, the Bush administration soon abandoned its position that it would not support some form of programmatic response to the rioting. What was an acceptable Republican position on urban poverty and mob violence?

Among the first GOP members of Congress to raise anxieties about rewarding the rioters was Kentucky Senator Mitch McConnell. Citing the previous criminal records of many arrestees in the Los Angeles disorders, he cautioned against financing any measure that would compensate "hooliganism and violence" ("Federal Assistance," 1992, p. S6987). McConnell deplored any action that would signal Uncle Sam's willingness to "reach into his wallet and bail you out" (p. S6987). He counseled his colleagues not to "reward random violence or incompetent government with federal financial windfalls" (p. S6987).

California Senator John Seymour took the reward-the-rioters rationale from words to action. The emergency appropriations bill, he advised his colleagues, would "reward rioters, looters, arsonists, and murderers" ("Federal Assistance," 1992, p. S6997). His personal predilections notwithstanding, Seymour could not afford to ignore the costly riot damages in his home state. His compromise was to propose an amendment restricting emergency federal aid to those who had not been arrested, convicted or charged with a riot-related crime. Another Republican, Texas Senator Phil Gramm, endorsed Seymour's measure and argued that the Treasury should never aid those "who burned and looted and killed" (p. S7006) in Los Angeles. That the amendment was supported by a large majority of the Senate is testament to the stigma its members attached to being associated with any effort to reward the rioters.

Thereafter, however, the theme failed to reappear in congressional discourse. Neither the enterprise zones measure encased in the ill-fated Tax Reform Act of 1992 nor the program clutched in the Budget Reconciliation Act of 1993 drew charges on Capitol Hill that new federal policies to aid cities might also aid those guilty of crimes such as arson, looting, or other mob actions. Once again, the fact that no further urban mob violence erupted in the months following the Los Angeles unrest eased the uncertainties of many congressional members. Without the pressures of new rioting, it became possible for them to support enterprise zones and empowerment communities. This situation marked a departure from the earlier Model Cities experience, in which there was no post-riot emergency appropriation to separate short-term riot response from longer-term social action.

When Model Cities was enacted in the middle of President Johnson's term, Congress had never before endorsed legislation specifically designed to address urban mob violence. This condition helps account for the fact that Model Cities was never clearly identified as a response to such cataclysmic events. It was billed, instead, as a gesture to reduce urban poverty—especially minority urban poverty. Even though private correspondence and public discourse in Congress clearly indicate the dual purposes of Model Cities, the precedent-setting nature of the program in American history could only have caused many members of the House and

Senate considerable discomfort. The twin pillars of the conditional violence rationale—the dangers of ignoring the riots and the dangers of
rewarding the rioters—represented a classic damned-if-you-do and damned-
if-you-don't conundrum.

Politicians at every level must contend with complex issues, all
options for which are likely to offend one or more sets of constituencies.
Yet few issues involve mass unlawful action often resulting in death and
always culminating in human injury, destruction of private property, and
the needed expenditure of millions of dollars in public funds. For this
reason, there was little in the background of members of Congress to
prepare them for the decisions they faced with regard to Model Cities or
subsequent War on Poverty legislation.

Faced with similar circumstances, the 102nd Congress, only a few
members of which were veterans of the War on Poverty, found itself
confronting the urban interracial mob violence conundrum more or less as
the 89th Congress had 26 years before. This time, however, the impending
threat of episodic rioting quickly subsided. This time, the social, political,
and economic status quo in the United States was vastly unlike that of 1966.
This time, videotaped documentation was available to corroborate charges
of police brutality and thereby sustain widespread public sympathy for the
immediate causes, if not the consequences, of mob violence. This time, the
issue of post-riot relief was partitioned from the larger issue of ameliorative
urban social policy. Was Congress's policy response—in this case the
EZEC programs—a sharp departure from Model Cities? I will explore this
matter in the next chapter. First however, I must contend with one final
element of the problem context.

Metaphorical Revisionism

One of the most intriguing aspects of urban interracial mob violence
in America is the language we create to express these events. Although
terms such as *riots, mayhem,* and *violence* are commonly employed by
journalists, public officials, and scholars, in the eyes of some observers
they all suffer from a certain indelicacy. Underlying their straightforward
denotation of destructiveness are their connotations of purposelessness or,
worse yet, willful—even retaliatory or wanton—actions for their own sake
(*rioting for fun and profit,* as Banfield, 1974, terms them). Some liberals

felt discomfort characterizing mob violence in those terms and believed— or wanted to believe—that there was deeper meaning and purpose when large numbers of primarily poor, black inner-city residents turned to arson, looting, robbery, vandalism, beating, wounding, and killing in reaction to a precipitating incident such as excessive use of force in a police arrest.[4] ¶ 179 Consequently, they sought to vest those activities with new meaning. As the 1960s progressed, members of the press, public officials, civil rights organizations, black militant groups, academicians, and liberal political figures increasingly referred to actions in terminology implying a more purposeful response to unfair circumstances. The sheer persistence of mob violence throughout the 1960s contributed to the impression among some observers that, rather than random, spontaneous outbursts, these events were part of an organized expression of outrage toward discrimination, inequality, and injustice. With the uncertainties that such claims raised, it was possible to pose a counterbalancing metaphor to the prevailing image of the violence as lawless, wanton, vengeful, and barbaric.

In the aftermath of the Los Angeles episode of 1992, however, there were few attempts in the popular media or among public officials to identify the actions as anything other than riots. On Capitol Hill, for example, one of the fairest and most judicious statements—that of black Congressman Craig Washington—condemned both the beating of Rodney King and the actions of those who participated in the mob violence after the court verdict was announced. Although deeply sympathetic to the rage of the crowds and the injustices symbolized by the King beating, Washington used terms such as *rioting* and *looting* with no apparent unease. Conservatives such as Senator Seymour described participants in the Los Angeles violence in terms such as "looters, arsonists and murders," and "hoodlums, hooligans and thugs" ("Federal Assistance," 1992, p. S6997). Apparently few attempts materialized to characterize mob violence in ameliorative words. Senator Christopher Dodd, for example, called the actions "unrest" and "discontent" (p. S6977). Congressman Henry Gonzalez referred to "unrest," "protest," and "disturbances" ("Introduction of Legislation," 1992, p. H3080). Only Representative Maxine Waters portrayed the events as "rebellion" ("Introduction of Legislation," 1992, p. H3080). Beyond these few references, the 1992 Los Angeles eruptions never achieved in metaphor the redemptive quality sought by liberals in response to the 1960s urban violence. Had outbursts spread 1960s-style to other cities over the spring and summer of 1992, a similar effort at emblematic transformation may have emerged. In the absence of such a pattern, it was difficult for most observers to conceive of the violence as a revolt or insurrection. Yet if ever in history there was persuasive evidence to support a claim of mob action based at least in part on legitimate grievances, it lie in the graphic scenes of the videotaped Rodney King beating. Never before had so many viewed with their own eyes a stark portrayal of the very kinds of official abuse alleged by black arrestees and witnesses for generations. Whatever the veracity of countless previous claims of police discrimination and brutality toward blacks over the years, the King incident could raise doubts about such charges in the future.

Although redemptive metaphors of the type common in the late 1960s failed to take root after the 1992 Los Angeles violence, a different type of metaphor emerged. It was suggested during House floor debate over federal emergency funding for Los Angeles and Chicago in May 1992. Congressman F. James Sensenbrenner, who supported emergency aid for Los Angeles, dismissed similar financing for the Windy City, arguing that flooding damage from the collapse of underground tunnel walls was due to municipal gross negligence. Referring to the tragedy as a manmade disaster, he proposed an amendment, which failed adoption, to forbid emergency federal aid to Chicago for any purposes related to the flooding incident. In doing so, Sensenbrenner sought to draw a distinction between damage in Los Angeles and Chicago, implying that the former was not due to manmade circumstances such as official neglect or misbehavior.

A week later, Senator Seymour reminded his colleagues that several natural disasters had befallen his home state of California, including the San Francisco earthquake of 1989, a disastrous neighborhood fire in Oakland, wildfires in the Los Angeles region, mud slides, and other events. All these had been declared emergencies or major disasters under the federal Stafford Disaster Relief and Emergency Assistance Act (Public Health and Welfare Act, 1989). In asking sympathy for his constituents, Seymour implied that the Los Angeles riots should be understood by his Senate colleagues as equally worthy of federal aid. As a Republican, he was in an especially uncomfortable position. Although many in his state expressed sympathy for Rodney King, few could excuse the record-breaking death and destruction in Los Angeles following acquittal of King's assailants. Seymour sought to tiptoe through the minefield of political reactions by supporting emergency aid for Los Angeles but attaching a provision restricting recipients to those who had not participated in the law breaking. In doing so, Seymour tried to distance himself from the necessity of appealing for aid under a program that historically has been employed to respond to naturally occurring events such as earthquakes, hurricanes, tornadoes, floods, forest fires, and the like. Yet, as he must have been aware, urban interracial mob violence—indeed, all forms of group violence—are primarily acts of commission. They are expressions of the human will rather than "acts of God" over which humans have essentially no control. As such, riots bring issues distinct from every other type of political or policy decision faced in Washington. How then, could Seymour avoid the appearance of treating manmade violence as yet another natural disaster while maintaining his support for federal disaster assistance? He did so by insisting that those who carried out the arson, looting, and other unlawful acts be separated from those who did not and that they be excluded from benefits. Thus, only those who had suffered loss of property and related harm but had not been among those causing the violence would be eligible for help. This placed disaster assistance recipients in Los Angeles in the same category as those receiving assistance after a natural event. In other words, Seymour succeeded in segregating the element of human commission from the damage and suffering caused by the fires, vandalism, and other destructive acts.

In this task, he was aided by his fellow conservative, Senator Phil Gramm. Gramm reminded the Senate that spending tax revenues in the wake of natural disasters involves no chance that human culpability will be rewarded. In Los Angeles-type calamities, on the other hand, where damage and death result from human commission, there is too great a risk that emergency aid will benefit those guilty of the crimes. Perhaps a bit uncomfortable about these issues, Illinois Senator Alan Dixon, who sought emergency aid for Chicago, agreed that neither Los Angeles nor Chicago had experienced a natural disaster.

But Seymour was not yet finished teasing out the finer distinctions in the issue of federal financing for disasters of nature versus disasters of human nature. In response to his question, Senator Barbara Mikulski confirmed that the Stafford Act was interpreted liberally by Federal Emergency Management Agency (FEMA) officials with regard to the matter of damage caused by threats other than fire. She was referring to wording in the law defining major disaster. *Major disaster* means any natural catas-trophe (including any hurricane, tornado, storm, high water, wind driven water, tidal wave, tsunami, earthquake, volcanic eruption, landslide, mud slide, snowstorm, or drought), "or, regardless of causes, any fire, flood or explosion" (Public Health and Welfare Act, § 2, 1989, p. 1116).

Eligibility to receive federal emergency assistance under provisions of the law includes any emergency created by fire, flood, or explosion, all three phenomena of which can be caused by natural events or by human actions. Presumably, because fire and water are elements of nature, any harm caused by them through acts of human commission was found by Congress to be sufficiently like harm growing out of natural disasters. Hence, Mikulski confirmed to Seymour that damage caused by fires at the hand of rioters, whether from Molotov cocktails, matches, lighters, torches, or other devices, would be covered by provisions of the Stafford Act. As for Los Angeles damage from sources other than fire, flood, or explosion (i.e., looting, vandalism, firearms discharges), Mikulski assured her col-league that the law would not distinguish between victims of fires in Los Angeles and victims of other damaging actions. Nowhere in the language of the Stafford Act is this issue clarified, however. Thus, it was that damage resulting from a distinctly nonnatural event—urban interracial mob vio-lence—was accommodated by the 102nd Congress in terms synonymous with those used to characterize meteorological, seismic, or hydrologic events.

More important than this episode on the Senate floor, however, is the peculiar precedent it sets: In the future, mob violence may be addressed in terms and procedures just like any natural disaster. (Ironically and tragi-cally, the Los Angeles area was to suffer wildfires and a severe earthquake within 2 years of the riots, adding further to the emergency assistance poured into the region.) In combination with the flooding damage due allegedly to official negligence in Chicago, the Los Angeles riots revealed what is perhaps a new capacity within the federal government for subtle revisionism in response to political expediency. As Republican California Congressman Campbell reminded his colleagues, national policy toward riot-related damages was once covered by the federal riot reinsurance program. His attempt to reinstate the program, which lapsed in 1984, recalled that official Washington had at one time felt property owners and businesses damaged by rioting were responsible for purchasing insurance (albeit federally subsidized) from the private sector. Congressional action in the wake of the Los Angeles riots, however, demonstrated that a barely perceptible shift occurred. A new metaphor has taken root, propagated in the rhetorical soil of Capitol Hill, but likely to be nurtured in the bureaucratic humus of government offices such as FEMA. In the 1960s, voices characterized rioting as a form of human protest, implying exculpation or absolution for those involved. The 1990s have effected an official reinterpretation of such events, quietly positioning acts of man in the same league as acts of God.

Notes

1. More than 20 years ago, Haar (1975) pointed out that place-based programs such as Model Cities could not overcome the "centrifugal forces of the metropolitan market" (pp. 259-260). Not only do such narrowly focused programs risk losing their political support, he notes, but confining their effects to poor neighborhoods overlooks the larger scale in which the underlying economic causes are at work.

2. It was no accident that, whereas the Model Cities program was attached to a bill including other urban and metropolitan programs, enterprise zones were buried in the Revenue Act of 1992 and empowerment zones were buried in the Omnibus Budget Reconciliation Act of 1993. This made it possible for any member of Congress to vote for either zones bill while insisting that support for the revenue or budgetary objectives was his or her intent. Because revenue or budgetary measures benefit more voters than the zones proposals, there was political safety in coupling the initiatives in this way.

3. The Kennedy-Hatch proposal would spend an additional $1.45 billion on existing programs to provide summer jobs and education for school-age youths and on Head Start for preschoolers. In addition, Hatch inserted President Bush's new Weed and Seed crime control proposal in the measure. This effort to extend what was supposed to be an emergency appropriations bill targeted to relief for Los Angeles, and later to Chicago, to all American cities was a clear sign that Kennedy recognized the short-term power of the danger-of-violence rationale. Beneath the surface was the implicit message that preemptive federal spending was merited to reduce the probability of additional mob violence in other cities.

4. In this case, *purpose* refers to violence as an expression of mass social protest with deep and persistent underlying causes triggered by some immediate action, usually by civil authorities. Therefore, terms such as *rebellions, revolts, civil disobedience,* and *insurrections* came into vogue. The Kerner Commission Report, for example, employed terms such as *riots* and *disorders* interchangeably. By summer of 1966, Vice President Hubert Humphrey was referring to ghetto *revolts*.

LIVERPOOL JOHN MOORES UNIVERSITY
LEARNING SERVICES

10

Urban Poverty, Interracial Mob Violence, and Federal Reaction

The Policy Context

T wice since the mid-1960s, the federal government has responded to urban interracial mob violence with unique, place-based national policies. Model Cities and urban enterprise zones-empowerment communities targeted eligibility, participation, and benefits in predominantly low-income city neighborhoods, often of the type spawning riots in the past.

The ways in which the problems of urban mob violence and poverty are defined—and the political atmosphere within which they are addressed—shifted significantly from the 1960s to the 1990s. Both the nature of urban poverty in America and the circumstances under which mob violence erupted in Los Angeles no longer lend themselves to the same interpretations that they had a quarter century ago. The urban loss of jobs, income, and tax base due to corporate capital flight, robotics and automation, deindustrialization, suburbanization, and the loss of middle-class social control and family stability in many poor urban neighborhoods together have conspired to shift congressional conceptions of the poverty problem from policy responsive in the 1960s to policy resistant in the 1990s. If poverty has become a tougher nut to crack in the 1990s though, the underlying role played by racial discrimination has been made manifest, given unprecedented, visually documented evidence of police brutality toward African American victims.

Also engendering sympathy in the Rodney King case was the restraint exercised in Los Angeles's minority communities after the videotape was released. Instead of exploding in spontaneous mayhem, as was often the pattern in the 1960s, residents of these areas maintained composure in anticipation of the outcome of the trial of the four police officers. The 102nd and 103rd Congresses were confronted with a somewhat divergent set of social circumstances than had arisen when the Model Cities legislation was considered. These new circumstances included a more policy-resistant strain of urban poverty and a more sympathetic rationale to account for the outbreak of mob violence.

It was not only the shifting problem context that presented Capitol Hill with a difficult dilemma in 1992. The quarter century had brought substantial reconfiguration of the political context within which the problem was framed and policy responses were crafted. Structurally, Congress was less prepared to react generously to the problems of urban poverty and mob violence. The congressional membership was more Republican and less interested in urban and Frostbelt issues, particularly those associated in the popular mind with liberal Democratic constituencies. Moreover, the diversity of issues and causes deemed appropriate for federal intervention had grown substantially since the 1960s—and with it, demands on the federal budget. Responses to urban mob violence and poverty had to compete with a much broader array of policies and programs, many of which benefited larger segments of society.

Even more complicated than structural shifts in the political context were altered conceptual patterns. Republicans were championing policies with little, if any, track record and therefore no legacy of failure to contend with. Democrats, on the other hand, could no longer look to a national ethos of moral responsibility propelled by the civil rights movement to advance their positions. The violence conundrum, which worked to the advantage of the urban poor in the 1960s, no longer did so in the 1990s. Signs of a continuation of urban rioting throughout America had contributed to the legitimacy of the danger-of-violence rationale for enacting Model Cities and diminished the momentum of the reward-the-rioters counterargument. In the wake of the 1992 Los Angeles riots, however, there were few indications that rioting would spread to other cities. The danger-of-violence admonition carried less weight (and the reward-the-rioters carried more) on Capitol Hill. This resulted in the enterprise zones-empowerment communities (EZEC) program, with more restrictive programmatic designs and more narrowly circumscribed benefits than Model Cities.

Yet the political context on Capitol Hill in 1992 and 1993 was not entirely antagonistic to the issues at hand. A decline in racially divisive discourse among members of Congress indemnified policy debates from the worst kinds of mischaracterizations heard in the 1960s. Metaphors on the House and Senate floors no longer implied that riots were rebellions against repressive conditions; instead, human commission itself was deemphasized by some members in deference to an alternative vision of riots as tantamount to natural disasters. A subtle shift in the conceptual definition suggested that federal policy toward rioting would now be addressed in terms not unlike those previously reserved for acts of God. Nonetheless, these more sympathetic trends in the 1990s could not offset the decidedly less favorable atmosphere in Congress toward alleviating conditions contributing to urban poverty and mob violence.

From the time of Martin Luther King, Jr., to that of Rodney King, congressional consideration of federal responses to urban rioting occurred within quite dissimilar problem and political contexts. What, then, did this imply for the policy outcomes? In the first section of this chapter, I compare and contrast the Model Cities and urban EZEC programs. Then I examine the bureaucratic transformation of urban EZECs under the Clinton administration. I then review the people-versus-place debate and argue that both programs were solidly place based. I follow with a discussion of the Clinton administration's efforts to subsume urban welfare policies in reconfigured national social welfare policy and the decidedly people-based emphasis therein. Finally, I reflect on emerging social welfare policies under the hegemony of the new Republican congressional majority and raise questions about implications for urban poverty and mob violence.

Model Cities and Urban EZECs:
A Comparison

Table 10.1 shows that urban EZECs' resources are concentrated in 71 communities, as opposed to the 150 local Model Cities programs. The challenge to Housing and Urban Development (HUD) staff of competent program management and evaluation should be considerably less daunting with EZECs than was the case with the Johnson-era measure.

Model Cities and urban EZECs also differ in terms of their authorized duration. Whereas the 1960s program was a demonstration with a life of 5 years for each model neighborhood, each enterprise community and empowerment zone is allowed up to 10 years of operation. A lengthier life span should allow the EZEC programs more time to prove themselves and a longer period to weather the attacks of an unfriendly intervening Congress or presidential administration.

On the other hand, urban EZECs' imposed funding mechanisms are considerably more complicated than that applied to Model Cities. The highly flexible supplemental funds under Model Cities, for example, heralded the later emergence of federal block grants, a most malleable funding device. Although urban EZECs include a block grant system and categorical grants similar in nature to those of Model Cities, they also employ tax relief measures. Tax credits and tax-exempt bonds, in turn, are incentives, the success of which are dependent on attracting private capital. The uncertainties associated with ostensibly risky inner-city investments may imperil the value of such subsidies. These conditions are further complicated by the fact that urban EZECs do not specify to local governments the assumption of a minimum nonfederal share of program costs as Model Cities did. This places a greater burden on localities to pledge higher contributions to compete with other applicants. Thus, localities are encouraged to propose local and state tax relief or direct subsidies, in-kind contributions, low-interest loans, capital facilities donations, and similar forms of participation. Perhaps having a greater financial stake in an EZEC will raise local commitment. But the complex diversity of funding sources necessary under EZECs seems to be a threat to overall program success.

Another basis of contrast between the two programs involves local oversight and management responsibilities. Each municipal participant in the Model Cities program was required to set up a city demonstration

TABLE 10.1 A Comparison of the Federal Model Cities and Urban EZEC
Programs

	Model Cities Program	*Urban EZEC Program*
No. of local program designees	150	71
Life of each local program	5-year demonstration	Up to 10 years
Federal funding mechanisms	Supplemental grants, categorical grants from other programs	Social services block grants, tax credit on wages paid to EZECs new employees, tax-exempt facilities bonds, categorical grants from other programs
Federal funding responsibilities	Supplemental grants = up to 80% of the costs of nonfederal programs; categorical grants = up to 80% of the nonfederal share could be paid from supplemental grants	No federal/nonfederal share specified; applicants encouraged to maximize local contributions to secure an EZECs designation
Local program operation and oversight	City Demonstration Agency, Model Neighborhood Board, municipal government	Municipal county, state governments, universities, medical centers, business corporations, nonprofit corporations
Local participation in decisions	Stresses "widespread citizen participation" from within the model neighborhood	Stresses that "affected community" be a "full partner" in "developing and implementing the plan"

SOURCES: Demonstration Cities and Metropolitan Development Act of 1966; Empowerment Zones,
Enterprise Communities, and Rural Development Investment Areas Act (1993); Hornbeck (1994).

agency (CDA) and a model neighborhood board. In the Model Cities
program, a complicated power split between city halls, CDAs, and model
neighborhood boards often resulted (Frieden & Kaplan, 1975). In urban
EZECs, local authorities are given greater latitude to design their own
coordination structures and procedures. Related to this matter is the role
of citizens, especially those living or working in the target area. Model
Cities emphasized widespread citizen participation from within the model

neighborhood. In some situations, complicated internal power struggles caused progress to go adrift (Frieden & Kaplan, 1975). Urban EZECs open the participatory door to a wider array of stakeholders, including state and local governments, nonprofit organizations, charitable groups, local businesses, and foundations. EZECs call for a full partnership for the affected community in preparing and carrying out the plan. Whether and to what extent localities will assign significant influence to those living or working in the EZEC remains to be seen. In one respect, less participation from insiders may allow others to move program development forward more expeditiously. In another respect, however, the experience of Rebuild L.A. has demonstrated the liabilities of giving too many interests a voice in program decision making. If everyone has his or her hands on the steering wheel, the organization can quickly lose its course and become immobilized.

The urban EZEC legislation targets a smaller number of cities over a longer time frame and offers a more complicated and uncertain financing system than Model Cities did. Urban EZECs elicit greater private sector participation than the Model Cities program did, allow local determination of organizational arrangements, and assign local influence on program decisions to a larger array of interests, including those living outside the designated program area. What accounts for these dissimilarities? Many appear related to the differing problem and political contexts within which Congress debated responses to the Los Angeles riots in 1992 and 1993. For example, with Republican and nonurban members in greater numbers, it became necessary to transform the distinctly urban Model Cities paradigm into a mixed urban-rural EZEC paradigm. It was also important to fund it with a combination of direct subsidies and tax incentives, the better to make peace with Republican wishes. Taking into account the federal deficit and greater competition for federal funds, fewer program designees had to be named. Similarly, resources are doled out over a longer time frame so as to reduce annual effects on the federal budget. Republicans favor more extensive private sector participation in local EZEC programs, another device for reducing federal expenditures. They also ensured that local program governance and influence would include not only those inside the target area boundaries but other city and regional interests as well. Without the moral authority of a national civil rights movement, the EZEC bill could not elicit from Congress the same concern over empowering urban minorities reflected earlier in the Model Cities program (the phrase empowerment zones notwithstanding). In the new climate, self-determination for zone residents takes a back seat to other objectives.

Program Implementation and Bureaucratic Transformations

Although variations in program design between Model Cities and EZECs merit attention, so do variations in the federal organizational setting of each program. Federal agency implementation of Model Cities brought about a number of challenges. One of these was resolving interagency conflicts necessary to bring about the concentration of diverse federal resources on individual model neighborhoods.[1] The same challenge now faces the EZEC initiative. In this regard, the Clinton administration seems to have absorbed some powerful lessons from the difficulties encountered by Model Cities' Washington Interagency Coordinating Council (WICC). President Clinton's Community Enterprise Board (CEB) has more direct lines of authority to the White House than the WICC had. Chaired by Vice President Al Gore, the CEB is made up of the secretaries of most of the cabinet agencies, including HUD and the Department of Agriculture. Also included are two vice chairs (both assistants to the president), the attorney general, the chair of the Council of Economic Advisers, the director of the Office of Management and Budget, and the heads of the Office of National Drug Control Policy, Environmental Protection Agency, and Small Business Administration. Formed by the president in September 1993, the CEB is charged with overseeing the designation of urban and rural EZEC participants. It is also required to bring cooperation among a host of cabinet agencies and to channel grants, loans, tax relief, technical assistance, and other resources from existing federal programs to each zone and community.

The WICC, on the other hand, was chaired by a subcabinet presidential appointee, a HUD assistant secretary. Its membership was also composed of sub-cabinet-level program managers, many of whom were career civil servants. As a result, the WICC never commanded the prestige or authority of the White House to the extent envisioned. The challenge of the CEB will come in maintaining the commitment of all members and in conveying down the chain of command in each participating federal agency that the White House expects cooperation. Cabinet legal counsels, for example, cannot be allowed to tie up progress when presidential discretion is clearly permitted. Should the CEB allow its responsibilities to filter down to agency underlings, its success will certainly be undermined. In the best of circumstances, federal interagency cooperation toward common goals is difficult to attain. Without unflagging White House support, the CEB may well suffer a fate similar to that of the WICC.

If interagency coordination is a critical element in the successful implementation of the EZEC program, a second point of intervention is the transformation of its scope and purposes as it works its way through HUD. Although the original enterprise zones concept—and to a lesser extent the EZEC legislation—emphasized economic development and employment growth in designated areas, the Clinton administration cobbled together a more complicated set of guidelines and policies for designing and operating local EZECs. Applicants for an EZEC designation are required to prepare and submit a strategic plan with a vision of the kind of community participants and residents desire in the future. The vision is to be based on four building blocks: neighborhood, balance, human scale, and restoration. According to HUD, good neighborhood building blocks include walkable streets, livable and affordable housing, accessible transportation, public spaces, civic facilities, and design principles such as an identifiable edge and center and linkages to the community and region (*Building Communities,* 1994).

The second building block calls for balancing economic development with physical and human development. These include a diversified job base with new businesses and protection of existing businesses, mixed-use projects that integrate the social and commercial centers of a neighborhood, a combination of car and transit opportunities, public and private support for individuals and families, and mixed housing types and ownership opportunities (*Building Communities,* 1994).

Human scale, the third building block, is defined as enhancing face-to-face interactions, pedestrian use, increased personal safety, and a sense of place. The final building block, restoration, urges applicants to restore cultural identity, physical history, and unique natural systems. Included are the target area's architecture, natural features, and art, along with its economic and social resources (*Building Communities,* 1994).

Where did the building blocks come from? If any were articulated on Capitol Hill, they are unrecorded. Perhaps not surprisingly, HUD identifies organizations with community-based planning experience such as the American Institute of Architects, the American Planning Association, the Urban Institute, and the National Civic League as among those which contributed to the vision statements (*Building Communities,* 1994). Much as the Johnson administration once sought Washington interest group support for Model Cities, the Clinton administration appears to have done so for EZECs. Yet implementation of the new urban program did not stop with building blocks.

HUD identifies four key principles as the selection criteria to be employed in awarding EZEC designations: economic opportunity, sustainable community development, community-based partnerships, and strategic vision for change. Economic opportunity includes promise of job, business, and entrepreneurship growth; increased economic stake in the target area among residents; involvement of residents in implementing the plan and in all phases of economic and community development; linkage between residents and employers in the region; job training and family support to become economically self-sufficient; ties between target area economic revitalization and the regional economy; and increases in lending and investment opportunities in the target areas (*Federal Register,* 1994).

Sustainable community development draws on a concept that has been advanced for several years by environmentalists seeking limits on physical development and growth subject to the capacity of natural resources such as air and water, open space, and plant and animal species to sustain these activities without further degradation. Vice President Gore's reputation as an environmentalist and his following among such interests helps account for its inclusion. Community-based partnerships called for participation in the EZEC program by all segments of the community. HUD has sought collaboration among nonprofit institutions, businesses, local and state governments, religious organizations, neighborhood groups, environmental advocates, and the like. A strategic vision for change should address community needs, be comprehensive in scope, indicate coordination mechanisms, describe a vision for the program, and stipulate goals and procedures for ongoing evaluation of the results (*Federal Register,* 1994).

HUD elaborates each of the four principles, giving a long list of more specific objectives, including providing affordable housing, seeking drug abuse abatement, providing equal opportunity, strengthening families, building confidence and skills among youth, protecting public health and the environment, involving residents in the EZEC program, creating jobs (both inside and outside the zone or community), and spurring investment in the zone or community. Other HUD objectives include creative and innovative activities; a process of revitalization linking economic, human, physical, and community development, as well as and other activities in a mutually reinforcing, synergistic way. Finally, HUD indicates no priorities among the four sets of criteria and instead stipulates that all carry equal weight in selection of awards (*Building Communities,* 1994).

However worthy these lofty aspirations are, they hazard the very real possibility that original EZEC core objectives such as reducing neighborhood unemployment will become clouded by a host of other purposes and ambitions. It is not clear whether HUD, the Department of Agriculture, and the CEB intend that these building blocks be complementary to essentials such as household self-sufficiency and poverty reduction or whether they carry equal weight in terms of desired program outcomes.

It is difficult to understand how a variegated collection of public, private, and nonprofit organizations and interests, working across multiple jurisdictions, attempting to pursue a complicated four-part vision according to four sets of multifaceted principles, will achieve results that significantly reduce target area poverty through increased household self-sufficiency. On the other hand, it takes little imagination at all, especially in view of the Model Cities experience, to see how misperceptions about program outcomes, based on lengthy laundry lists of criteria and objectives, could result in management delays, groups working at cross-purposes, misallocation of resources, and disappointment over program outcomes. Add to these dilemmas the extreme difficulty of coordinating individual federal assistance programs across program and agency boundaries, both in Washington and in federal regional offices. Bringing these multiple federal resources (in concert with diverse state, local, public, and private resources) to bear on individual target areas could rival the Allied invasion of Europe in 1944 for sheer daunting complexity.

Model Cities demonstrated several things, one of which is that multiple goal attainment is difficult to measure effectively and convincingly;

the larger the list of goals, the more elusive the job of measurement and the more likely that the attainment of one goal will come at the expense of others. In effect, Congress and HUD promised too much to too many people and interests. Model neighborhoods were charged with building and renovating housing, creating and running day care centers and after-school programs for youths, subsidizing health clinics, fighting crime and drugs, stimulating new businesses, training and employing residents, and a host of other functions. The result was that most programs simply could not deliver enough to all who expected results. With the added burdens of an indifferent, and later hostile, Nixon White House, flagging fiscal support from Congress, a weak WICC, and a short program life, Model Cities could not sustain the multiple promises nor minister to the diverse expectations of all the intended beneficiaries. These problems led to the eventual abandonment of the program. The ultimate outcome was that few, if any, of the original target neighborhoods showed evidence of permanently reducing unemployment, poverty, or household welfare dependency, much less arresting physical and economic neighborhood decline (Wood, 1990).

A lack of clear and simple priorities and program focus, combined with complicated lines of federal decision making, may be the headiest challenge for the EZEC program in the years ahead. Paradoxically, the original enterprise zones objectives were relatively simple to understand: to raise people out of poverty through employment development, job training, and selected neighborhood investments by offering incentives to private investors, businesses, and entrepreneurs. The something-in-it-for-everyone identity being created within the urban EZEC program is sure to set up misperceptions and give ammunition to those who will later seek to discredit it. Secretary Henry Cisneros and the Clinton administration (or their successors) would be well-advised to dispense with the extensive list of building blocks, principles, and objectives quietly. Instead, two or three measurable objectives related to reducing poverty and increasing employment and self-sufficiency among low-income households should be articulated. Each EZEC participant should understand that continuation of support for the local program will be based on consistent, measurable progress toward achievement of these goals. The underlying principle should be a simple one: increasing household self-sufficiency within low-income target urban neighborhoods.

Urban Policy and the
People-Versus-Place Debate

I have tried to show how Congress reacted to major episodes of urban mob violence in two periods of time: the mid-1960s and the early 1990s. I have noted that, in both periods, many congressional members assumed that poverty was a critical underlying factor in accounting for the causes of these tragedies. In both periods, Congress responded by enacting place-based programs—Model Cities and urban EZECs—targeted to predominantly poor urban areas. Although the two programs are comparable on the surface, substantially dissimilar problem and political contexts account for substantially dissimilar policy designs. Moreover, bureaucratic implementation of urban EZECs has revealed that some, although not all, lessons from the Model Cities experience have been remembered.

As important as these issues are, the larger questions associated with place-based urban policies must not be dismissed. Has Washington been misdirected in attempts to intervene in structurally rooted conditions as complex as poverty through place-based programs such as Model Cities and urban EZECs? A corollary questions the wisdom of attempts to lessen the likelihood of urban mob violence through place-based policies that target poverty. In this section, I review the debate over place- and people-based urban policies, concluding that urban EZECs offer little promise of significantly enhancing economic self-sufficiency among target area populations. Nor is the program likely to have much effect on the probability of urban mob violence occurring.

In the interim between the explosions in Watts and in South-Central Los Angeles, an important debate emerged in Washington over appropriate federal policies toward America's cities and its poor. It was crystallized in *Urban America in the Eighties* (President's Commission for a National Agenda for the Eighties, 1980), the final report of a commission appointed by President Carter that was released at the end of his administration. The central theme the report poses is the "people-versus-place" issue (Glickman, 1983; Hicks, 1983). The report raises the question whether public subsidies to alleviate poverty and urban decline could most effectively be targeted at households, on the one hand, or at cities, neighborhoods, and downtowns, on the other. Helping people, the report maintains, implies subsidies regardless of the location of the intended beneficiaries. Examples include supplementing household income, training workers, improving and expanding educational services, and providing health care insurance. Helping places, the report points out, encompasses measures that restrict assistance to recipients, purposes, and programs in a geographically delimited area. In particular, place-based policies have historically centered on subventions for housing, community development, infrastructure, mass transit, urban fiscal improvements, small business development, and job creation, especially in downtowns and urban neighborhoods.

Both people and place approaches have been a part of Washington's urban policy arsenal for many years. Nonetheless, the report argues that subsidizing places risks bonding the poor to inner cities with steadily diminishing opportunities (President's Commission for a National Agenda for the Eighties, 1980). Because continued assistance is so often limited to continued residency in the target area, those who benefit—or hope to benefit—must themselves remain living in such places. Only by doing so can they hope to become beneficiaries of policies that offer geographically based assistance for day care, employment training, rental assistance, food stamps, small business development, home improvements, public housing, and neighborhood infrastructure.

The commission raised doubts about how efficiently and effectively federal place-based programs reach the alleged beneficiaries (President's Commission for a National Agenda for the Eighties, 1980). The benefits of tax deductions and credits to builders and investors in downtown development, for example, rarely trickle down to those living in poverty; when they do, the results are often a few temporary construction jobs and low-paying employment as hotel maids, janitors, and parking attendants. On the other hand, subsidizing households offers people the chance for migration to more robust job opportunities in the suburbs or in other regions of the nation. With effective job training and education and temporary food, housing, and health care support, the report argues, the poor will have a better chance to achieve self-sufficiency.

The people versus place debate cast a shadow on then-existing urban place-based programs such as Urban Homesteading, Neighborhood Housing Services, Section 8 Housing, General Revenue Sharing, Community Development Block Grants, and Urban Development Action Grants. Indeed, subsequent to the commission's report, several of these programs were abolished, threatened, or diminished by the Reagan administration.

What does the people versus place debate imply for the urban EZEC program? The urban EZEC program, much like Model Cities before it, is a place-based initiative that assumes that geographically restricted, temporary assistance to poor neighborhoods can significantly reduce poverty and increase household self-sufficiency. Both programs were posited on the belief that, given a short-term helping hand, poor households and small businesses could reintegrate themselves into urban, metropolitan, and regional economies. Tragically, however, such premises fall victim to the fallacy that it is possible to rebuild families and communities from the inside out. Noble experiment that it was, Model Cities did not—and could not—reconstruct neighborhood economies. In that regard, even the most Olympian of neighborhood social and economic successes could not be sustained in the absence of continued, and probably perpetual, public subsidies. The Model Cities demonstration clarified that the American public will not tolerate such subsidies indefinitely.

To argue that EZECs will suffer a similar fate is not to claim that the two programs are identically designed or intentioned.[2] The EZEC program is a creature of its times. The 1993 legislation centered on only one real goal: jobs for low-income people. It was founded on the assumption that direct and indirect federal, state, and local subsidies and regulatory relief will attract and maintain the presence of viable employers in designated low-income urban neighborhoods. Some local programs will realize greater success than others. Some will be able to point to visible enhancements to the neighborhood such as new buildings, day care centers, and happy workers. But the overall success of previous attempts to attract and maintain new employers in minority and low-income neighborhoods in central cities—in short, to rebuild inner-city economies—does not give room for much optimism. Changes in product lines, market areas, or corporate leadership, together with vandalism, pilferage, and robberies, too often conspire to terminate these admirable efforts. In some cases, employers cannot find needed skills or education levels in the immediate neighborhood or in other poor neighborhoods nearby. The higher costs of insurance for urban businesses can elevate operating expenses. The cessation of subsidy mechanisms may also play a role. In any event, the result is that too few of those most in need of jobs in targeted low-income neighborhoods succeed in gaining and retaining steady employment.

Moreover, American businesses are more mobile than ever before. The old-fashioned notion that companies, once established, are committed and benevolent creatures of the community is an enduring image in American cultural history. If recent experience has taught us anything, however, it is that many such firms are quite prepared to relocate to the suburbs, to other regions, indeed to other nations, when and if circumstances dictate.

Of course, nonprofit institutions such as community development corporations can make modest inroads. With a different set of expectations and a different regimen than profit-based corporations, these organizations are usually more committed to neighborhood social and economic goals. But their markets and profits tend to be somewhat more tenuous because they are usually rooted locally. Rarely do they demonstrate the ability to achieve self-sufficiency, instead requiring continual public or corporate subsidies. Although they may succeed in modestly redistributing income within neighborhoods and central cities, they are rarely able to create new income by drawing capital from within their region, much less from larger market areas. As such, nonprofit organizations can play a role in EZEC programs, but the record to date gives little optimism that by themselves they can reverse the cycle of poverty in poor neighborhoods.

To argue that the urban EZEC program is unlikely to transform neighborhood or urban economies is not to insist that it will have no short-term or localized benefits. Relatively small numbers of the poor will complete job-training programs and find and hold employment with EZEC businesses. Some will remain with these companies, and others will use their training and experience as a stepping stone to better jobs elsewhere in the city or even in the suburbs. Some, especially the young, will seek higher education and training. Just as Model Cities had its individual success stories, so will EZECs.

But EZECs will also prove that cities thrive and prosper when and if their economies are interrelated with regional, national, and international flows of capital, investment patterns, creation of markets, and distributional profiles of products and services. The urban poor can benefit from such conditions when and if their skills, educational levels, and work habits match the requirements of employers physically accessible to them. Suppliers of goods and services to these businesses also benefit—for example, carryout restaurants, mobile lunch wagons and snack canteens, hair parlors, and dry cleaners. But without a strong and enduring linkage to regional, if not national and international, markets, subsidy-induced, place-based business development programs are constructed like a house of cards. They cannot exist in a vacuum indefinitely, regardless of corporate good intentions or local political will.

People-Based Approaches: Embedded Urban Policy and Mobility Strategies

The Clinton administration has struggled to formulate a strategy to address urban poverty. But the word *urban* seldom passes the lips of the president or members of his cabinet, who recognize the public unpopularity of cities and the poor in the mid-1990s. A *Washington Post* survey shortly after the 1992 Los Angeles riots found that 70% of respondents believed that American urban problems are essentially insurmountable (Morin, 1992). Clinton has sought to embed his urban policy in a broader campaign of social and economic justice such as a national health care plan and national welfare reform.

Embedded Urban Policy

Recognizing that public identification with cities and the poor among nonurban voters is hazardous politically, Clinton's has sought wider public support for his proposals. His ill-fated health care plan was majoritarian in its reach, if not its grasp. His successfully enacted crime bill promises assistance to community law enforcement units, not only in cities but in smaller communities as well. The Brady Bill, requiring handguns to be registered, and Clinton's efforts to ban the sale of assault weapons both have powerful urban implications but are portrayed as beneficial to all citizens. His welfare assistance proposals, languishing on Capitol Hill in early 1996, are directed not only to poor urbanites but also to the poor in all locations of the nation.

Since the brief flurry of activity in the wake of the Los Angeles riots, there has been little enthusiasm in Washington for instituting major initiatives identified as urban in intent (DeParle, 1994). Rebuilding downtowns, revitalizing neighborhoods, constructing significant numbers of low- and moderate-income housing, reforming city school systems—these have been low-profile issues since the Reagan presidency. The Clinton administration has been well advised by organizations such as the nonpartisan Urban Institute to embed urban-directed measures in larger social strategies that offer benefits to nonurban constituents as well (*Confronting the Nation's Urban Crisis,* 1992). The fundamental truth about these people-based programs is that, unlike place-based programs, they are not designed to confine their beneficiaries to areas in which prospects for rising out of poverty are severely limited (Skocpol, 1991). Although not prohibiting households from remaining in or near poor neighborhoods, neither do they condition the receipt of social services on continued residence in those areas. Instead, they are meant to provide families both geographic and socioeconomic mobility—to better housing, public education, jobs, social services, and the like. Under truly national programs, people have greater freedom to migrate to better opportunities with less fear of losing government-subsidized benefits.

In a political realm in which place-based policies have less room, the fate of the nation's welfare program, perhaps more so than other policies, foretells of the future of embedded urban policy. In 1996, Democrats and Republicans on Capitol Hill are struggling over several issues of welfare reform:

- Decentralization—states may be given increased latitude to fashion their own terms of eligibility, benefits, and levels of support.
- Workfare—recipients of welfare assistance may be expected to work in public or private sector jobs at least part time or to undergo job training followed by such employment.
- Job-production subsidies—To increase the chances for workfare, private sector employers may be offered subsidies or other incentives if they hire (or hire and train) welfare recipients. Public sector jobs with states and governments under subsidies may also be included.
- Duration of benefits—a time limit on the receipt of welfare benefits may be imposed by the federal government or state governments may be authorized to impose their own deadlines. Afterward, welfare recipients would no longer be eligible for support unless they find and hold a job.
- Entitlement—to continue welfare as a legal right available to all whose income qualifies them or to limit eligibility to those who are not able-bodied may be left to states to decide.

- Additional federal support for workfare may include child care arrangements and transportation to and from work sites.
- Penalties may be included that disqualify recipients for benefits if they bear out-of-wedlock children or that offer no additional benefits if they have more such children. (Pear, 1995)

The centerpiece of welfare reform at this point is household self-sufficiency through employment. Implied are the notions of work for anyone who wants it and perpetual welfare for no one who is unwilling to work. Whether such policies will ultimately reduce household dependency, much less discourage illegitimate births and school dropouts, remains to be seen. They do raise an interesting issue related to crime in general and mob violence in particular. If significant numbers of the urban poor could lose their subsidized job if convicted of a crime (with no chance of receiving welfare assistance), would that condition be a sufficient disincentive for participation in muggings, drug distribution, looting, or arson? Could it discourage illegal mob behavior? Could workfare, in other words, become a new method of social control?

One positive externality associated with a workfare policy could be additional support to state and local governments and private and nonprofit firms for carrying out socially beneficial tasks. Workfare jobs could include service in a day care center, assistance to law enforcement officers, rehabilitation of housing, remediation of toxic and hazardous waste, assistance to health care professionals, and other activities contributing to human welfare or environmental quality. Worrisome "disbenefits" are associated with workfare policies. For example, what effects will such policies have on the supply of and demand for nonsubsidized workers and jobs? If incentives are created to hire the poor, will those whose incomes place them above poverty status be penalized? If employers receive tax or other subsidies for hiring and training the poor, will this create an unfair disadvantage for those who must compete for the same jobs without the federal government behind them?

Mobility Strategies

If embedded, non-place-specific programs are the backbone of Clinton's domestic social policies, where does this leave HUD? Many of its current

programs are place-based, often limiting eligibility and benefits to municipalities or subareas. Many of these efforts were intended to improve
downtowns, neighborhoods, public facilities, and other locations or to
enhance housing stocks or housing opportunities in such places. Secretary
Cisneros, recognizing flagging support across the nation for such policies,
has decided not to put all his eggs in the placed-based programs basket.
Instead, HUD offers a series of people-based initiatives centered on enhancing mobility opportunities for poor households.

Among them is HUD's Moving to Opportunity for Fair Housing
research demonstration. Housing authorities in Boston, Baltimore, Chicago, Los Angeles, and New York receive special funding to offer Section
8 rental vouchers or certificates to households in high-poverty neighborhoods. Chosen by lottery from residents of public and assisted housing,
participating households are given training and assistance in finding suitable housing in low-poverty neighborhoods. This pattern is in stark contrast to existing housing programs for the poor, which tend to concentrate
in areas with high rates of poverty. A major issue is to examine household
progress out of poverty toward self-sufficiency in the years after the move.
Originally enacted by Congress in the waning days of the Bush administration, Moving to Opportunity has received enthusiastic support from
Cisneros (Gallagher, 1994). Enhancing household mobility out of poor
areas is a primary goal in at least two other HUD policy proposals. Both
of these would provide housing search assistance and more flexible regulations to encourage eligible poor families to move to areas with lower
concentrations of poverty in cities or suburbs (*Residential Mobility Programs,* 1994).

Aside from providing subsidies, training, and search assistance to poor
households, HUD has also increased its enforcement of antidiscrimination
housing laws. Since its enactment after the assassination of Martin Luther
King, Jr., the Fair Housing Act has been amended to assist not only racial
minorities but also families with children and people with disabilities.
Cisneros has placed greater emphasis than his predecessors on removing
barriers to housing access. One hope is that accessibility to rental and sales
housing in suburban communities will increase among low- and moderate-
income households seeking to leave central cities.

For all the attention it has engendered, there is nothing really new
about the concept of household mobility. Even during the height of Presi-

dent Johnson's War on Poverty, initiatives such as Model Cities and the Community Action Program were labeled by critics as efforts to "gild the ghettos." Some detractors called for programs that would "disperse the ghetto" to urban and suburban neighborhoods with fewer problems (Downs, 1968, 1973). Among other concerns, critics expressed doubts that ghetto enrichment programs could succeed in breaking the cycle of dependency among poor households or that poor central city neighborhoods could ever achieve other than perpetual government subsidization (Kain & Persky, 1969)

The foundation of ghetto dispersal strategies was erected in the late 1960s. The *spatial mismatch hypothesis,* as it came to be known, argues that a mismatch exists between the residential location of most poor blacks in central city neighborhoods and available job opportunities appropriate to their skill levels, which increasingly have appeared in the suburbs (Kain & Persky, 1969). Racial discrimination in suburban housing markets, the hypothesis maintains, keeps poor blacks in inner cities, whereas employers are decentralized to increasingly remote sites in metropolitan areas. Furthermore, low automobile ownership rates among urban blacks and lack of information about suburban opportunities hinder their ability to commute to such opportunities. Although some researchers find only partial merit in the spatial mismatch explanations for urban black poverty (Jencks & Mayer, 1990), a school of thought has grown up that has found greater support since the people-versus-place debate flourished in the early and mid-1980s (Ihlanfeldt, 1994; Jargowsky & Bane, 1991; Kain, 1992; Kasarda, 1989; McGeary, 1990). For example, in response to the L.A. riots in 1992, the Urban Institute recommended that mobility policies be adopted by the federal government (*Confronting the Nation's Urban Crisis,* 1992).

Further support for mobility strategies has grown out of encouraging results in HUD's Gautreaux Assisted Housing Program (Broder, 1994; Page, 1994). Initiated in response to a lawsuit challenging Chicago's largely racially segregated public housing stock, Gautreaux is HUD's attempt to remedy the problem. It involves counseling of African American public housing applicants and tenants to assist them in finding suburban market-rate rental housing, using Section 8 certificates to cover part of the rent. Counseling continues after tenants are relocated. Results to date have found that Gautreaux participants are more likely to achieve higher income, as well as employment and educational status, than control groups

(Davis, 1993; Rosenbaum & Popkin, 1991). A similar court-imposed
remedy is underway in Yonkers, New York, under HUD's Enhanced
Section 8 Outreach Program (Berger, 1994).

Cisneros and HUD officials have gone to great lengths to clarify that
the Clinton administration is hardly turning its back on its historic mission
to revitalize central cities and to reduce the conditions of urban poverty.
They emphasize giving greater choice to the poor to reside wherever they
desire. But they also realize that many of the nation's urban mayors and
minority interest groups are not enthusiastic about policies that threaten to
deconcentrate minority political strength, diminish minority institutions
such as church congregations, or undermine consumer demand for products and services of minority-based businesses. What is remarkable about
mobility and other nonplace federal programs today is that, until recently,
no presidential administration (and few members of Congress) would have
dared to favor policies that hazard such results. Such is the collective effect
of past place-based policies perceived to have failed; the people-versus-
place debate; the ascendancy of Republican, suburban, and Sunbelt interests in Congress; and an American electorate with few ties to urban
constituents.

In the short range, mobility policies are unlikely to turn significantly
over the social and racial makeup of concentrated pockets of poverty
households in urban neighborhoods. Because the number of participants
in HUD mobility programs is small, their effect on the neighborhoods they
leave behind, as well as those they relocate to, is likely to be modest. More
important are the long-term effects of mobility strategies. Assuming that
current HUD initiatives increase upward social mobility through geographical mobility and raise the level of household self-sufficiency, will
such programs continue to deliver the same outcomes? Will suburban
incumbent residents and politicians mount exclusionary actions to thwart
such measures? Will suburban communities reach a perceived "tipping
point," beyond which the addition of minority households will be actively
resisted? Or will another wave of white flight be initiated similar to those
occurring in the 1950s in response to black immigration to urban neighborhoods and in the 1970s and 1980s in response to middle-class black
migration to suburban communities (Gale, 1987)? Given past suburban
reactions to school desegregation, subsidized housing, and other policies
that threatened to introduce poor and minority households to middle-class

communities, substantial expansion of mobility programs seems politically unsustainable. Therefore an urban policy strategy, mobility is necessary but not sufficient (Hughes & Sternburg, 1992; McGeary, 1990).

If public and political perceptions are important to program survival, there is an even darker side to mobility strategies insofar as rhetorical abuse is concerned. If rioting is most likely to occur in neighborhoods with high concentrations of minority urban poor, it follows that deconcentrating such populations may reduce the propensity for mob violence. Dispersal could be labeled a *divide and conquer* strategy by critics, however, especially by skilled demagogues. Regardless of HUD protestations about fostering choice for poor families, mobility strategies could soon take on the identity of ghetto busters, feeding suspicions that a white plan exists to break up minority communities as a vehicle for social control. The federal Urban Renewal and Interstate Highway programs, as well as private market activities that spur gentrification, have elicited similar accusations (Gale, 1987).

Setting these detractions aside, what if expanded mobility programs succeed in achieving meaningful deconcentration of poverty in some urban neighborhoods over the longer term? Will this leave further housing abandonment and neighborhood decline in their wake (Hughes & Sternburg, 1992)? The risk of such results will vary in degree from city to city, of course, but it seems much more likely that other outcomes will prevail. These outcomes include the following:

- Succession—High-poverty neighborhoods vacated by participants in mobility programs will attract new poverty households (perhaps African Americans or other minorities) in the classic American succession pattern.
- Gentrification—Instead of attracting more poor families, neighborhoods where appealing architectural styles, historic character, pleasing views, or other attributes exist will attract childless middle-class households.
- Urban redevelopment—Neighborhoods located convenient to the central business district, waterfronts, or other "healthy" enclaves will undergo large-scale rehabilitation and redevelopment for hotels, offices, specialty retail, and up-market residential uses.

In cities such as Cleveland, Detroit, and St. Louis, with large areas of underpopulated and abandoned neighborhoods, mobility programs could exacerbate an already dire situation. Succession, gentrification, or urban redevelopment could not offset the effect of higher vacancies and rates of abandonment. In cities such as Denver, Boston, and Seattle, with lower proportions of such neighborhoods, mobility programs might result in any one or some combination of these three processes. Other variables include the level of middle-class demand for urban residential space, accessibility of mobility target neighborhoods to employment centers and other city attractions, and the willingness and experience of city administrations to broker public-private partnerships. The likelihood of succession outcomes will be determined in considerable degree by the continued immigration of legal and illegal aliens to the United States and to a lesser extent by the rural-to-urban migration of poor whites and African Americans. The likelihood of significant gentrification outcomes is diminished somewhat by the passing of the post-World War II baby boom bulge, leaving proportionally smaller age cohorts at the household formation stage to seek residence in such neighborhoods. Nonetheless, the Urban Institute recommends that gentrification be encouraged by governments as a means of offsetting the rise in income segregation between cities and suburbs (*Confronting the Nation's Urban Crisis,* 1992). Both gentrification and urban redevelopment outcomes may be hindered by forces such as telecommunications technology, suburban employment growth, urban crime, and poor urban public schools, all of which draw households and businesses away from central cities and to the metropolitan areas.

Congress, the Urban Poor, and Mob Violence

In the fall 1994 national elections, Republicans succeeded in gaining a majority in the House of Representatives after decades of Democratic dominance. President Clinton lost the Democratic majority in both houses of Congress that he had enjoyed for 2 years. With Republicans now in many positions of power on Capitol Hill, Congress is attempting to reduce the federal deficit through massive reductions in spending. Not since the New Deal has domestic social policy undergone such momentous rewriting. The result is likely to be substantial reductions in federal employment and the downsizing, elimination, or consolidation of individual cabinet departments, units within departments, and independent agencies. The president can veto any bill enacted by Congress, but he will use his discretion and accept most Republican-supported actions as a necessary political price.

HUD Secretary Cisneros, hoping to head off a Republican campaign to eliminate his department, has proposed reducing HUD's workforce by more than one third by the year 2000 ("Housing Dept. Plans," 1995). In addition, Cisneros proposes eliminating 21 HUD field offices and combining all current HUD programs into three large block grant programs, one each for affordable housing development, community development, and housing rent certificates (*Reinvention Blueprint,* 1994). Block grant recipient communities would be given greater latitude to develop their own priorities, procedures, and mechanisms for spending funds so long as certain federal principles (e.g., federal fair housing laws, support for homeless and disabled people) are met. Incentives in the form of additional grant support would be offered to communities that perform most effectively. Disincentives through reduced grant funding would be imposed on those that perform poorly. Public housing projects would be converted to mixed-income residential enclaves, with some tenants paying market rates and others receiving various levels of rent assistance. Increased geographical mobility of poor households would be encouraged through additional Section 8 certificates and vouchers. Other cabinet departments that serve cities and the poor, such as the Departments of Commerce, Labor, Education, and Health and Human Services, have also been singled out for substantial budget reductions.

Clearly, the federal government is headed toward a future in which astonishingly diminished resources will be expected to produce more effective and efficient outcomes for the poor, especially for the urban poor. Further decentralization of power from Washington and the federal agencies to state and local governments is likely. Place-based policies will not be eliminated and central cities will not be abandoned, but priority on assisting poor households to achieve independence and self-sufficiency, whether in central city or in suburban neighborhoods, is probable. Reductions in welfare rolls and increases in employment among the poor appear imminent.

In one respect, the Republican revolution is a measure of the depletion of new ideas among liberals after more than 50 years of social policy experimentation. So weary have many voters become of crime, drugs, poverty, interracial violence, unemployment, school dropouts, children born out of wedlock, multigenerational welfare dependency, and other social ills that voters' reserves of tolerance have finally been overdrawn. At the heart of the new conservatism in Washington are traditional values of independence, individual initiative, thrift, personal industriousness, volunteerism, and states rights. Unfortunately, many Americans have become so desperate for fundamental change that almost anything appearing to be the antithesis of the status quo is likely to find a modicum of support. But contrarian policies do not guarantee a just society.

Societal strategies that stimulate growth in personal responsibility, social mobility, and economic self-sufficiency certainly deserve experimentation. But wholesale restructuring of domestic social welfare systems in the absence of careful testing—and in pursuit of reelection in 1996— may simply substitute one set of human problems for another. Cutting budgets, reducing federal payrolls, narrowing program eligibility rules, and eliminating social programs could lead to increases in infant mortality, unemployment, homelessness, crime, and drug addiction. Could such actions also lead to new outbursts of rioting in the nation's inner cities?

• Urban Mob Violence in the Future

Both Model Cities and EZECs were enacted in direct response to major episodes of urban mob violence. Model Cities did not succeed in significantly reducing the conditions that it was believed contributed to mob violence in inner-city neighborhoods. Urban EZECs came into being as the most recent policy response to such tragedies. Only time will tell if EZECs will prevent further rioting. Few people are banking on any consistent pattern of salvation. Yet whenever the next flare-up of urban unrest occurs, the federal government once again will be called on to enact programs to curb such disasters by ameliorating their underlying causal factors.

Conservatives can point out that the worst period of rioting in the nation occurred during the latter 1960s, when African Americans in particular had more positive signs from Washington than at any other time perhaps since Emancipation and the end of the Civil War. Since 1969, however, Republican presidents have been in the White House for 20 of the past 26 years; from Richard Nixon's inauguration to mid-1995, major urban riots in America occurred only twice in Miami and once in Los Angeles. To explain the epidemic of civil unrest arising from 1964 to 1968, some observers advanced the *theory of the rising level of expectations.* With the War on Poverty in full swing and the civil rights movement achieving unprecedented reforms for African Americans, frustration grew over successes not yet realized. The more imminent new reforms seemed, the more intolerable became the status quo. The weight of hopelessness lifted by the successes of a Lyndon Johnson or a Martin Luther King, Jr., permitted some blacks to imagine a truly color-blind society. And although hope brought restraint among some, it unleashed anger and outrage among others. Thus, liberal reforms were accompanied by more, not less, mob violence.

Under such reasoning, one might expect the conservative trend in Washington and the nation to discourage urban mob violence. As it becomes increasingly clear to African Americans and other urban minorities that Washington intends even less assistance in fighting poverty than before, it would seem that hopelessness exceeding even that of the Reagan years will take root. Under such circumstances, little advantage in stimulating unrest will be perceived by potential rioters and comparative peace will reign in poor neighborhoods.

To some observers, this scenario might be reassuring. Yet to accept it may be to lull oneself into a kind of policy somnambulance. Most episodes of urban mob violence unfold as the result of a single incident in which an African American is arrested, beaten, or killed by one or more police officers. Often, although not always, these events occur in hot weather and are fed by rumors. There is little evidence that national social or political currents impel the catastrophes of collective violence. Neither the Harlem riots of 1936 or 1943, nor the Detroit riot of 1943, nor the Miami riots of 1980 and 1989, nor the Los Angeles riot of 1992 could be said to have taken place in a national climate of civil rights advancements or welfare reform. That the most concentrated period of urban mob violence in U.S. history coincided with the most ambitious period of minority advancement does not in any way obviate the potential for individual outbursts to occur in entirely dissimilar times. Although the epidemic character of urban unrest during those times may have been related to nationally current ideas or events, the capacity for isolated, catastrophic incidents of mob violence to appear in the years ahead leaves little room for retreat into a fool's paradise. The simple truth is that more interracial violence will erupt in cities and the seams of superficial quietude will be torn once again. Police and the National Guard will be mobilized, mayors and governors will be tested, and, ultimately, federal authorities will be asked to finance the cleanup, revisit the problem, and find solutions. Once again, Congress will react to symptoms, rather than causes. Perhaps another place-based program will be applied to the wound. But until politicians and the people realize that a fundamental restructuring of both the responsibilities and the rewards of life in a capitalistic, pluralistic, democratic society is necessary, the threat of continued urban mob violence will be with us. Unless and until those who are better off among us accept that interracial and interclass peace is most likely to come through a fundamental redistribution of wealth and income, the nation's cities will be our battlegrounds again and again.

Notes

1. So troublesome were the disjunctures between federal agencies, Frieden and Kaplan (1975) note, that it became impossible for the Model Cities program to achieve its goals. Indeed, these authors insist, the federal bureaucracy was the central problem. Levin (1987) unleashes unusual vitriol in his conclusions about HUD staff, accusing them of smugness and excessive careerism.

2. Others have explored the implications of Model Cities for the urban EZEC program. See Hetzel (1994) and Rubin (1994).

References

Advisory Council on Intergovernmental Relations. (1993). *Significant features of fiscal federation* (vol. 2).

Aftermath of the disturbance in Los Angeles. (1992, May 7). *Congressional Record,* 102nd Congress, 2nd Session, pp. H3081-H3086.

American cities in danger of blowing up in flames. (1992, May 12). *Congressional Record,* 102nd Congress, 2nd Session, p. H3097.

Anderson, M. (1964). *The federal bulldozer.* New York: McGraw-Hill.

Anthony, E. (1970). *Picking up the gun: A report on the Black Panthers.* New York: Deal.

Ayres, B. D., Jr. (1991, May 7). Street unrest flares again in capital. *New York Times,* p. 18.

Banfield, E. C. (1974). *The unheavenly city revisited.* Boston: Little, Brown.

Beamish, R. (1992, May 9). Bush pledges help for the cities. *Portland Press Herald,* p. 1.

Beauregard, R. (1989). Space, time, and economic restructuring. In R. Beauregard (Ed.), *Economic restructuring and political response* (pp. 209-240). Newbury Park, CA: Sage.

Beauregard, R. A. (1993). *Voices of decline: The postwar fate of U.S. cities.* Oxford, UK: Basil Blackwell.

Bennet, J. (1994, March 13). Hope starts to bloom in the motor city that no longer builds many cars. *New York Times,* p. 11.

Berger, J. (1994, June 12). A minority program is slowly trying to get a foot in the all-white door. *New York Times,* p. 19.

Booz, Allen, & Hamilton. (1971). *Study of the concentration and dispersion impact of the Model Cities program* (Vol. 1). Washington, DC: Department of Housing and Urban Development.

Bornet, V. D. (1983). *The presidency of Lyndon B. Johnson.* Lawrence: University of Kansas Press.

Bradbury, K., Downs, A., & Small, K. A. (1982). *Urban decline and the future of American cities.* Washington, DC: Brookings Institution.

Broder, D. (1993, April 11). Bradley, Cisneros trade ideas on urban revival. *Maine Sunday Telegram,* p. 7C.

Broder, D. (1994, June 10). Public housing strategy would depopulate "the projects" one family at a time. *Sun-Sentinel,* p. 23A.

Brown, R. M. (1969). Historical patterns of violence in America. In H. D. Graham & T. R. Gurr (Eds.), *Violence in America: Historical and comparative perspectives* (pp. 43-80). New York: Signet.

Browning, R. P., Marshall, D. R., & Tabb, D. H. (1984). *Protest is not enough: A theory of political incorporation.* Berkeley: University of California Press.

Brownstein, R. (1992, May 5). Bush, Clinton match revival ideas. *Portland Press Herald,* p. 2.

Building communities: Together. Guidebook: Strategic planning. (1994). Washington, DC: Department of Housing and Urban Development.

Button, J. W. (1978). *Black violence: Political impact of the 1960s riots.* Princeton, NJ: Princeton University Press.

Califano, J. A., Jr. (1991). *The triumph and tragedy of Lyndon Johnson: The White House years.* New York: Simon & Schuster.

Carter tells of his fears. (1993, March 5). *Portland Press Herald,* p. 2.

Cities bill passes main test. (1966, October 16). *New York Times,* p. IV, 2.

Civil Rights Act of 1968, Title VIII Fair Housing, Pub. L. No. 90-284, § 804-805, 82 Stat. 81-90 (1968).

Cohen, J., & Murphy, W. S. (1966). *Burn, baby, burn.* New York: Dutton.

Community Capital Partnership Act of 1993. (1993, April 30). *Congressional Record,* 103rd Congress, 1st Session, pp. S5290-S5306.

Conference report on H.R. 11, Revenue Act of 1992. (1992, October 5). *Congressional Record,* 102nd Congress, 2nd Session, pp. H11633-H11655.

Confronting the nation's urban crisis: From Watts (1965) to South Central Los Angeles. (1992). Washington, DC: Urban Institute.

Congress adjourns, leaving Bush with dilemma on tax bill. (1992, October 10). *Boston Globe,* p. 4.

Congressional Record. (1966a, August 19). 89th Congress, 2nd Session, pp. 20060-20070.

Congressional Record. (1966b, October 14). 89th Congress, 2nd Session, pp. 26922-27043.

Congressional Record. (1966c, October 20). 89th Congress, 2nd Session, p. 28139.

Congressional Record. (1992a, May 14). 102nd Congress, 2nd Session, pp. H3266-H3675.

Congressional Record. (1992b, August 12). 102nd Congress, 2nd Session, pp. H8131-H8134.

Congressional Record. (1992c, September 29). 102nd Congress, 2nd Session, p. S15604.

Cowden, R. (1995, February). Power to the zones. *Planning,* pp. 8-10.

Darden, J. T., Hill, R. C., Thomas, J., & Thomas, R. (1987). *Detroit: Race and uneven development.* Philadelphia: Temple University Press.

Davis, M. (1993). The Gautreaux Assisted Housing program. In G. T. Kingsley & M. A. Turner (Eds.), *Housing markets and residential mobility* (pp. 21-32). Washington, DC: Urban Institute Press.

Demonstration cities. (1966, October 13). *New York Times,* p. 44.

Demonstration Cities and Metropolitan Development Act of 1966, Pub. L. No. 89-754. (Title I, Comprehensive City Demonstration Programs). (1966). *U.S. Code Congressional and Administrative News,* Vol. 3, pp. 3999-4014.

DeParle, J. (1994, March 27). Clinton wages a quiet but energetic war against poverty. *New York Times,* p. 6.

Dommel, P. R. (1985). The evolution of community development policy. *Journal of the American Planning Association, 51*, 476-478.

Downs, A. (1968). Alternative futures for the American ghetto. *Daedalus, 97*, 1331-1378.

Downs, A. (1973). *Opening up the suburbs: An urban strategy for America.* New Haven, CT: Yale University Press.

Drake, S., & Cayton, H. R. (1962). *Black metropolis: A study of Negro life in a northern city* (Vol. 1.). New York: Harper.

Eaton, W. J. (1992, June 19). Congress passes emergency urban aid bill. *Portland Press Herald,* pp. 1A, 8A.

Elmi, A. H. N., & Mikelsons, M. (1991). *Housing discrimination study: Replications of 1977 study measures with current data* (HUD-1329-PDR). Washington, DC: Department of Housing and Urban Development.

Emergency time-sensitive assistance for American youth. (1992, May 14). *Congressional Record,* 102nd Congress, 2nd Session, p. S6684.

Empowerment Zones, Enterprise Communities, and Rural Development Investment Areas Act, Pub. L. No. 103-66, 107 Stat. 543-558 (1993).

Enterprise zones. (1992, August 11). *Congressional Record,* 102nd Congress, 2nd Session, pp. S12222-S12250.

Fager, C. E. (1974). *Selma, 1965.* New York: Scribner.

Fainstein, S. S., Fainstein, N. I., Hill, R. C., Judd, D. R., & Smith, M. P. (1983). *Restructuring the city: The political economy of urban redevelopment.* London: Longman.

Fairclough, A. (1987). *To redeem the soul of America.* Athens: University of Georgia Press.

Feagin, J. R., & Hahn, H. (1973). *Ghetto revolts: The politics of violence in American cities.* New York: Macmillan.

Federal assistance in addressing urban crisis. (1992, May 20). *Congressional Record,* 102nd Congress, 2nd Session, pp. S6974-S7017.

Federal Register, 58(11), January 18, 1994, pp. 2701-2707.

Fogelson, R. M. (1968). Violence as protest. In R. H. Connery (Ed.), *Urban riots: Violence and social change* (pp. 146-158). New York: Columbia University Press.

Fogelson, R. M. (1969a). *The Los Angeles riots.* New York: Arno Press and New York Times.

Fogelson, R. M. (Ed.). (1969b). *Mass violence in America: The Los Angeles riots.* New York: Arno Press and New York Times.

Fogelson, R. M. (1969c). White on black: A critique of the McCone Commission Report on the Los Angeles riots. In R. Fogelson (Ed.), *The Los Angeles riots* (pp. 111-144). New York: Arno Press and New York Times.

Fossett, J. W., & Nathan, R. P. (1981). The prospects for urban revival. In R. Bahl (Ed.), *Urban government finance: Emerging trends* (pp. 63-104). Beverly Hills, CA: Sage.

Frey, W. H. (1993). People in places: Demographic trends in urban America. In J. Sommers & D. A. Hicks (Eds.), *Rediscovering urban America: Perspectives on the 1980s* (pp. 3.23-3.106). Washington, DC: Department of Housing and Urban Development.

Frieden, B. J., & Kaplan, M. (1975). *The politics of neglect: Urban aid from Model Cities to revenue sharing.* Cambridge: MIT Press.

Gale, D. (1987). *Washington, DC: Inner-city revitalization and minority suburbanization.* Philadelphia: Temple University Press.

Gallagher, M. L. (1994, July). HUD's geography of opportunity. *Planning, 60,* 12-13.

Gallup, G. H. (1972). *The Gallup Poll, 1935-1971.* New York: Random House.

Galster, G. (1993). Polarization, place and race. *North Carolina Law Review, 71,* 1421-1462.

Glickman, N. J. (1983). National urban policy in an age of economic austerity. In D. A. Hicks & N. J. Glickman (Eds.), *Transition to the 21st century: Prospect and policies for economic and urban-regional transformation* (pp. 301-343). Greenwich, CT: JAI.

Haar, C. M. (1975). *Between the idea and the reality: A study in the origin, fate and legacy of the Model Cities program.* Boston: Little, Brown.

Haas, L. J. (1992, May 23). Cities may have a long, long wait. *National Journal, 24,* 1243.

Hackett, G. (1989, January 30). All of us are in trouble: Miami's riots die down but racial tensions linger. *Newsweek,* p. 36.

Hager, G., & Cloud, D. S. (1993, August 7). Democrats tie their fate to Clinton's budget bill. *Congressional Quarterly,* p. 2122.

Harris, L. (1971, October 4). The Harris Survey: Black animosities found increasing. *Washington Post,* p. A8.

Hetzel, O. J. (1994). Some historical lessons for implementing the Clinton administration's empowerment zones and enterprise communities program: Experience from the Model Cities program. *Urban Lawyer, 26,* 63-81.

Hicks, D. A. (1983). Urban and economic adjustment to the postindustrial era. In D. A. Hicks & N. J. Glickman (Eds.), *Transition to the 21st century: Prospect and policies for economic and urban-regional transformation* (pp. 345-370). Greenwich, CT: JAI.

Hicks, D. A., & Rees, J. (1993). Cities and beyond: A new look at the nation's urban economy. In J. Sommers & D. A. Hicks (Eds.), *Rediscovering urban America: Perspectives on the 1980s* (pp. 2.2-2.125). Washington, DC: Department of Housing and Urban Development.

Hill, R. C., & Negrey, C. (1987). Deindustrialization in the great lakes. *Urban Affairs Quarterly, 22,* 580-597.

Hornbeck, J. F. (1994). *Empowerment zones/enterprise communities: Can a federal policy affect local economic development?* Washington, DC: Congressional Research Service.

Housing dept. plans a big cut in its work force. (1995, January 8). *New York Times,* p. 12.

Hughes, M. A., & Sternburg, H. (1992). *The new metropolitan reality: Where the rubber meets the road in antipoverty policy.* Washington, DC: Urban Institute.

Humphrey warns of slum revolts. (1966, July 19). *New York Times,* p. 19.

Hunter, M. (1966a, August 18). Katzenbach warns of new city riots. *New York Times,* p. 31.

Hunter, M. (1966b, August 19). Money called no panacea for urban problems. *New York Times,* p. 18.

Hunter, M. (1966c, August 20). Senate approves Johnson slum aid. *New York Times,* pp. 1, 26.

Huxtable, A. L. (1966, February 21). Toward excellence in urban redesign. *New York Times,* p. 12.

Ihlanfeldt, K. (1994). The spatial mismatch between jobs and resident locations within urban areas. *Cityscape, 1,* 219-244.

Insiders win, cities lose in Senate tax handout. (1992, August 7). *USA Today,* p. 6A.

Introduction of legislation to provide emergency loan guarantee assistance to the city of Los Angeles. (1992, May 7). *Congressional Record,* 102nd Congress, 2nd Session, p. H3080.

Janowitz, M. (1968) Patterns of collective racial violence. In H. D. Graham & T. R. Gurr (Eds.), *Violence in America: Historical and comparative perspectives* (pp. 393-422). New York: New American Library.

Janson, D. (1966, July 16). Troops restoring order in Chicago Negro ghetto; 2 dead, 57 hurt in rioting. *New York Times,* p. 1.

Jargowsky, P., & Bane, M. J. (1991). Ghetto poverty in the United States, 1970-1980. In C. Jencks & P. E. Peterson (Eds.), *The urban underclass* (pp. 253-273). Washington, DC: Brookings Institution.

Jencks, C., & Mayer, S. E. (1990). Residential segregation, job proximity, and black job opportunities. In L. E. Lynn, Jr., & M. G. H. McGeary (Eds.), *Inner city poverty in the United States* (pp. 187-222). Washington, DC: National Academy Press.

Jones, S. J. (1969). *The government riots of Los Angeles, June 1943*. Saratoga, CA: R & E Research Associates.

Kain, J. F. (1968). Housing desegregation, Negro employment, and metropolitan decentralization. *Quarterly Journal of Economics, 82*, 175-197.

Kain, J. (1992). The spatial mismatch hypothesis: Three decades later. *Housing Policy Debate, 3*, 371-460.

Kain, J., & Persky, J. J. (1969). Alternatives to the gilded ghetto. *Public Interest, 14*, 74-87.

Kantor, P., & David, S. (1988). *The dependent city: The changing political economy of urban America*. Glenview, IL: Scott, Foresman.

Kasarda, J. D. (1983, March). Entry-level jobs, mobility, and urban minority unemployment. *Urban Affairs Quarterly, 19*, 21-40.

Kasarda, J. D. (1989). Urban industrial transition and the underclass. In W. J. Wilson (Ed.), *The ghetto underclass: Social science perspectives* (pp. 26-47). Newbury Park, CA: Sage.

Kasarda, J. D. (1993). Inner city poverty and economic access. In J. Sommer & D. A. Hicks (Eds.), *Rediscovering urban America: Perspectives on the 1980s* (pp. 4.1-4.60). Washington, DC: Department of Housing and Urban Development.

Kearns, D. (1976). *Lyndon Johnson and the American dream*. New York: Harper & Row.

Knight, R. V., & Gappert, G. (Eds.). (1989). *Cities in a global society*. Newbury Park, CA: Sage.

Koritz, D. (1991, June). Restructuring or destructuring? Deindustrialization in two industrial heartland cities. *Urban Affairs Quarterly, 26*, 497-511.

Kranish, M. (1992, May 17). Mayors lead DC rally for more urban aid. *Boston Sunday Globe*, p. 2.

Kranish, M., & Gosselin, P. G. (1992, May 17). Aides split over Bush urban plan. *Boston Sunday Globe*, p. 19.

Krauss, C. (1991, May 8). Latin immigrants in capital find unrest a sad tie to past. *New York Times*, p. 1.

Krauss, C. (1992, July 3). House passes aid plan for inner cities. *New York Times*, p. A10.

L.A. and the economics of urban unrest. (1994). In *The Taubman Center Annual Report* (pp. 4-5). Cambridge, MA: Harvard University, John F. Kennedy School of Government.

Lemann, N. (1991). *The promised land: The great black migration and how it changed America*. New York: Knopf.

Lemann, N. (1994, January 9). Rebuilding the ghetto doesn't work. *New York Times Magazine*, pp. 15-60.

Levin, M. R. (1987). *Planning in government*. Washington, DC: Planners Press.

Levitan, S., & Johnson, C. M. (1986). Did the Great Society and subsequent initiatives work? In M. Kaplan & P. L. Cuciti (Eds.), *The Great Society and its legacy* (pp. 73-90). Durham, NC: Duke University Press.

Logan, J. R., & Swanstrom, T. (Eds.). (1990). *Beyond the city limits: Urban policy and economic restructuring in comparative perspective*. Philadelphia: Temple University Press.

Marine, G. (1969). *The Black Panthers.* New York: New America Library.

Marshall Kaplan, Gans, and Kahn. (1970a). *The Model Cities program: A comparative analysis of the planning process in eleven cities.* Washington, DC: Department of Housing and Urban Development.

Marshall Kaplan, Gans, and Kahn. (1970b). *The Model Cities program: The planning process in Atlanta, Seattle, and Dayton.* Washington, DC: Department of Housing and Urban Development.

Marshall Kaplan, Gans, and Kahn. (1973a). *The Model Cities program: A comparative analysis of participating cities process, product, performance and prediction.* Washington, DC: Department of Housing and Urban Development.

Marshall Kaplan, Gans, and Kahn. (1973b). *The Model Cities program: Ten model cities—A comparative analysis of second round planning years.* Washington, DC: Department of Housing and Urban Development.

Marx, G. T. (1971). Civil disorder and the agents of social control. In D. Boesel & P. H. Rossi (Eds.), *Cities under siege: An anatomy of the ghetto riots, 1964-1968* (pp. 157-184). New York: Basic Books.

Massey, D., & Denton, M. A. (1993). *American apartheid: Segregation and the making of the underclass.* Cambridge, MA: Harvard University Press.

McGeary, M. G. H. (1990). Ghetto poverty and federal policies and programs. In L. E. Lynn, Jr., & M. G. H. McGeary (Eds.), *Inner city poverty in the United States* (pp. 223-252). Washington, DC: National Academy Press.

Meier, A., & Rudwick, E. (1971). Black violence in the 20th century: A study in rhetoric and retaliation. In H. D. Graham & T. R. Gurr (Eds.), *Violence in America: Historical and comparative perspectives* (pp. 380-392). New York: New American Library.

Mollenkopf, J. (1983). *The contested city.* Princeton, NJ: Princeton University Press.

Morin, R. (1992, May 18-24). Plenty of blame to go around. *Washington Post National Weekly Edition,* p. 37.

Moynihan, D. P. (1969). *Maximum feasible misunderstanding.* New York: Free Press.

The muddled Model Cities model. (1992, July 3). *New York Times,* p. A24.

Mufson, S., & Crenshaw, A. B. (1992, August 5). Tax legislation gains new life. *Washington Post,* pp. 1, 17.

Negroes riot in Cleveland. (1966, July 19). *New York Times,* p. 1.

Negro youths attack police and loot Chicago stores. (1966, July 13). *New York Times,* p. 1.

The Newark riot. (1971). In D. Boesel & P. H. Rossi (Eds.), *Cities under siege: An anatomy of the ghetto riots, 1964-1968* (pp. 19-66). New York: Basic Books.

News release: Statement of Henry Cisneros on announcement of empowerment zones and enterprise communities. (1994). Washington, DC: Department of Housing and Urban Development.

No demonstration cities. (1966, May 15). *New York Times,* p. 14.

Noyelle, T. J., & Stanback, T. M., Jr. (1983). *The economic transformation of American cities.* Totowa, NJ: Rowman & Allanheld.

Page, C. (1994, April 29). Color-blind view of housing opportunities denies cruel realities of suburbs. *Sun-Sentinel,* p. 19A.

Pear, R. (1995, July 30). Clinton and Dole bidding to break welfare impasse. *New York Times,* pp. 1, 12.

Peterson, P. (1985). Introduction: Technology, race and urban policy. In P. Peterson (Ed.), *The new urban reality* (pp. 1-29). Washington, DC: Brookings Institution.

Pomfret, J. D. (1966, July 24). Johnson asserts riots by Negroes impede reforms. *New York Times,* p. 213.

Portes, A., & Stepick, A. (1993). *City on the edge: The transformation of Miami.* Berkeley: University of California Press.

Powledge, F. (1970). *Model city: A test of American liberalism—One town's efforts to rebuild itself.* New York: Simon & Schuster.

President's Commission for a National Agenda for the Eighties. (1980). *Urban America in the eighties.* Englewood Cliffs, NJ: Prentice Hall.

President signs civil rights bill; pleads for calm. (1968, April 12). *New York Times,* p. 1.

Property damage exceeds $100 million. (1980). *Facts on File Yearbook, 40,* 382.

Public Health and Welfare Act, Title 42, Pub. L. No. 100-707. U.S.C. (pp. 1112-1135). (1989).

Reese, L. A. (1993, March). Decision rules in local economic development. *Urban Affairs Quarterly, 28,* 501-513.

Reinvention blueprint. (1994). Washington, DC: Department of Housing and Urban Development.

Reischauer, R. (1986). Fiscal federalism in the 1980s: Dismantling or rationalizing the Great Society. In M. Kaplan & P. L. Cuciti (Eds.), *The Great Society and its legacy* (pp. 179-197). Durham, NC: Duke University Press.

Remarks of President George Bush to community leaders in Los Angeles. (1992, May 28). *Congressional Record,* 102nd Congress, 2nd Session, p. E1563.

Report of the National Advisory Commission on Civil Disorders. (1968). New York: Bantam.

Republicans care about urban problems. (1992, May 13). *Congressional Record,* 102nd Congress, 2nd Session, pp. H3174-H3177.

Request to consider on today or any day thereafter H.R. 5132, Dire Emergency Supplemental Appropriations Act, 1992, for disaster assistance to meet urgent needs because of calamities such as those which occurred in Los Angeles and Chicago. (1992, May 12). *Congressional Record,* 102nd Congress, 2nd Session, pp. H3174-H3177.

Residential mobility programs. (1994, September). *Urban Policy Brief,* pp. 1, 5-6.

Rezendes, M. (1993, April 18). LA calm after two guilty verdicts. *Boston Sunday Globe,* p. 1.

Riot Reinsurance Act of 1992. (1992, June 22). *Congressional Record,* 102nd Congress, 2nd Session, p. E1935.

Rivera, C. (1992, August 6). U.S. riot relief plan offers little new, critics say. *Los Angeles Times,* pp. A1, A12.

Rosenbaum, J. E., & Popkin, S. J. (1991). Employment and earnings of low-income blacks who move to middle-class suburbs. In C. Jencks & P. E. Peterson (Eds.), *The urban underclass* (pp. 342-353). Washington, DC: Brookings Institution.

Rossi, P. H. (1978). Issues in the evaluation of human services delivery. *Evaluation Quarterly, 2,* 573-599.

Rubin, K. (1981). *Community development in Gainesville.* Cambridge, MA: Harvard University, John F. Kennedy School of Government.

Rubin, M. M. (1994). Can reorchestration of historical themes reinvent government? A case study of the Empowerment Zones and Enterprise Communities Act of 1993. *Public Administration Review, 54,* 161-169.

Rugaber, W. (1966a, July 22). Cleveland police wound Negro mother, 3 children. *New York Times,* p. 1.

Rugaber, W. (1966b, July 20). Negro killed in Cleveland; guard called in new riots. *New York Times,* p. 1.

Rugaber, W. (1966c, July 21). Trouble persists in Hough section. *New York Times,* p. 1.

Sawers, L., & Tabb, W. K. (1984). *Sunbelt/snowbelt: Urban development and regional restructuring.* New York: Oxford University Press.

Schussheim, M. J. (1974). *A modest commitment to cities.* Lexington, MA: Lexington Books.

Sears, D. O., & McConahay, J. B. (1973). *The politics of violence: The new urban blacks and the Watts riots.* Boston: Houghton Mifflin.

Semple, R. B., Jr. (1966a, October 18). Conferees agree on slum measure. *New York Times,* p. 18.

Semple, R. B., Jr. (1966b, October 15). Demonstration Cities bill passed by House, 178-141. *New York Times,* p. 1.

Semple, R. B., Jr. (1966c, October 11). Demonstration Cities bill urged by 22 top business executives. *New York Times,* p. 26.

Semple, R. B., Jr. (1966d, June 29). House unit votes slum aid program. *New York Times,* p. 26.

Semple, R. B., Jr. (1966e, June 23). Johnson is backed on cities program. *New York Times,* p. 1.

Semple, R. B., Jr. (1966f, March 3). Lindsay backs U.S. slum plan but calls amount inadequate. *New York Times,* p. 21.

Semple, R. B., Jr. (1966g, February 5). Mayors cautious on new plan. *New York Times,* p. 17.

Semple, R. B., Jr. (1966h, May 24). Move on to save bill to end slums. *New York Times,* p. 63.

Semple, R. B., Jr. (1966i, January 26). New urban plan would use federal funds to eradicate slums. *New York Times,* p. 18.

Semple, R. B., Jr. (1966j, October 19). Senate approves attack on slums. *New York Times,* p. 1.

Semple, R. B., Jr. (1966k, November 4). Signing of Model Cities bill ends long struggle to keep it alive. *New York Times,* p. 1.

Semple, R. B., Jr. (1966l, March 7). Slum plan stirs concern in House. *New York Times,* p. 1.

Semple, R. B., Jr. (1966m, June 16). U.S. told suburbs must help cities. *New York Times,* p. 1.

Sense of the Senate resolution on enterprise zones. (1993, June 24). *Congressional Record,* 103rd Congress, 1st Session, pp. S7963-S7965.

Serrano, R. A., & Wilkinson, T. (1992, April 29). All 4 in King beating acquitted. *Los Angeles Times,* p. 1.

Setting a firm course for American domestic policy. (1992, May 13). *Congressional Record,* 102nd Congress, 2nd Session, pp. H3225-H3234.

Silver, A. A. (1968). Official interpretations of racial riots. In R. H. Connery (Ed.), *Urban riots: Violence and social change* (pp. 146-158). New York: Columbia University Press.

Sims, C. (1994, May 22). Who said Los Angeles could be rebuilt in a day? *New York Times,* p. 5.

Sitkoff, H. (1978). Racial militancy and interracial violence in the Second World War. In R. Lane & J. J. Turner (Eds.), *Riot, rout and tumult: Readings in American social and political violence* (pp. 307-326). Westport, CT: Greenwood.

Sitkoff, H. (1981). *The struggle for black equality, 1954-1980.* New York: Hill & Wang.

Skocpol, T. (1991). Targeting within universalism: Politically viable policies to combat poverty in the United States. In C. Jencks & P. Peterson (Eds.), *The urban underclass* (pp. 411-436). Washington, DC: Brookings Institution.

Slum aid slashed by Senate panel. (1966, July 27). *New York Times,* p. 24.

Smith, M. P., & Feagin, J. R. (1987). *The capitalist city.* New York: Basil Blackwell.

Smith, T. (1966, July 19). Mayor sees E. Harlem's gaiety and woe. *New York Times,* p. 1.

Sobel, L. (Ed.). (1967). *Civil rights, 1960-66.* New York: Facts on File.

Sparks, R. M. (1992, September 9). Urban aid: An election year confrontation. *Area Development,* pp. 27-28.

Stanback, T. M., Jr., Bearse, P. J., Noyelle, T., & Karasek, R. A. (1981). *Services: The new economy.* Totowa, NJ: Allanheld, Osmun.

Stanback, T. M., Jr., & Noyelle, T. (1982) *Cities in transition.* Totowa, NJ: Allanheld, Osmun.

Sundquist, J., & Davis, D. (1969). *Making federalism work.* Washington, DC: Brookings Institution.

Tax Enterprise Zones Act. (1992, September 25). *Congressional Record,* 102nd Congress, 2nd Session, pp. S15024-S15042.

The toll from the riot. (1992, August 6). *USA Today,* p. 9A.

To rescue the cities. (1966, October 6). *New York Times,* p. 46.

Turner, M. A. (1992). Discrimination in urban housing markets: Lessons from fair housing audits. *Housing Policy Debate, 3,* 185-215.

Turner, M. A., Fix, M., & Struyk, R. J. (1991). *Opportunities denied, opportunities diminished: Racial discrimination in housing.* Washington, DC: Department of Housing and Urban Development.

20 years after the Kerner Commission: The need for a new civil rights agenda. (1990). Washington, DC: Government Printing Office.

Unanimous consent agreement—Senate Resolution 324. (1992, July 2). *Congressional Record,* 102nd Congress, 2nd Session, pp. S9664-S9682.

Understanding the riots: Los Angeles before and after the Rodney King case. (1992). Los Angeles: Los Angeles Times.

Urban aid. (1992, December 19). *Congressional Quarterly,* pp. 3867-3868.

VanBuskirk, P. (1972). *The resurrection of an American city.* Cambridge, MA: Schenkman.

Wade, R. (1971). Violence in the cities: A historical view. In D. Boesel & P. H. Rossi (Eds.), *Cities under siege: An anatomy of the ghetto riots, 1964-1968* (pp. 277-296). New York: Basic Books.

Wade, R. C. (1959). *The urban frontier.* Chicago: University of Chicago Press.

Wallace, M. (1978). The uses of violence in American history. In R. Lane & J. J. Turner, Jr. (Eds.), *Riot, rout, and tumult: Readings in American social and political violence* (pp. 10-27). Westport, CT: Greenwood.

Washnis, G. J. (1974). *Community development strategies: Case studies of major model cities.* London: Praeger.

Wehrwein, A. C. (1966, July 14). Negroes in Chicago clash with police; 2 hurt by gunfire. *New York Times,* p. 1.

White, T. H. (1965). *The making of the president, 1964.* New York: Athaneum.

Wikstrom, G., Jr. (1974). *Municipal response to urban riots.* San Francisco: R & E Research Associates.

Wilson, W. J. (1980). *The declining significance of race: Blacks and changing American institutions* (2nd ed.). Chicago: University of Chicago Press.

Wilson, W. J. (1987). *The truly disadvantaged: The inner city, the underclass and public policy.* Chicago: University of Chicago Press.

Wong, K. K. (1990). *City choices: Education and housing.* Albany: State University of New York Press.

Wood, R. C. (1990). Model Cities: What went wrong—The program or its critics? In N. Carmon (Ed.), *Neighbourhood policy and programmes* (pp. 61-73). London: Macmillan.

Woodlawn Organization. (1970). *Woodlawn's Model Cities plan: A demonstration of citizen responsibility.* Northbrook, IL: Whitehall.

Woods, F. J. (1982). *The Model Cities program in perspective: The San Antonio, Texas experience.* Washington, DC: Government Printing Office.

Workers' Family Protection Act. (1992, September 15). *Congressional Record,* 102nd Congress, 2nd Session, pp. S13484-S13495.

Index

About the Author

Dennis E. Gale is Henry D. Epstein Distinguished Professor of Urban and Regional Planning. He is Chairman of the Department of Urban and Regional Planning in the College of Urban and Public Affairs at Florida Atlantic University. Formerly, he was Director of Planning and Management Research at the Urban Institute in Washington, DC. He later acted as Professor of Urban and Regional Planning and Director of the Center for Washington Area Studies at George Washington University, where he shared the Dilthey Prize for research in 1984. He holds graduate degrees from Boston University, Harvard University, and the University of Pennsylvania; he earned a doctorate from George Washington University. He is the author of *Neighborhood Revitalization and the Postindustrial City* (1984) and *Washington, D.C.: Inner City Revitalization and Minority Suburbanization* (1987). His interests in federal urban policy were advanced during his participation in the White House Conference on Balanced National Growth and Economic Development in 1978. Other research and teaching interests include housing and community development, historic preservation, and growth management and land use policy.